Services and Business Process Reengineering

Series Editors

Nabendu Chaki, Department of Computer Science and Engineering, University of Calcutta, Kolkata, India

Agostino Cortesi, DAIS, Ca' Foscari University, Venice, Italy

The book series aims at bringing together valuable and novel scientific contributions that address the critical issues of software services and business processes reengineering, providing innovative ideas, methodologies, technologies and platforms that have an impact in this diverse and fast-changing research community in academia and industry.

The areas to be covered are

- Service Design
- Deployment of Services on Cloud and Edge Computing Platform
- Web Services
- IoT Services
- Requirements Engineering for Software Services
- Privacy in Software Services
- Business Process Management
- Business Process Redesign
- Software Design and Process Autonomy
- Security as a Service
- IoT Services and Privacy
- Business Analytics and Autonomic Software Management
- Service Reengineering
- Business Applications and Service Planning
- Policy Based Software Development
- Software Analysis and Verification
- Enterprise Architecture

The series serves as a qualified repository for collecting and promoting state-of-the art research trends in the broad area of software services and business processes reengineering in the context of enterprise scenarios. The series will include monographs, edited volumes and selected proceedings.

More information about this series at http://www.springer.com/series/16135

Santosh Kumar Pani · Manjusha Pandey
Editors

Internet of Things: Enabling Technologies, Security and Social Implications

Editors
Santosh Kumar Pani
Kalinga Institute of Industrial Technology
Deemed to be University
Bhubaneswar, India

Manjusha Pandey
Kalinga Institute of Industrial Technology
Deemed to be University
Bhubaneswar, India

ISSN 2524-5503　　　　　　ISSN 2524-5511　(electronic)
Services and Business Process Reengineering
ISBN 978-981-15-8620-0　　ISBN 978-981-15-8621-7　(eBook)
https://doi.org/10.1007/978-981-15-8621-7

© The Editor(s) (if applicable) and The Author(s), under exclusive license to Springer Nature Singapore Pte Ltd. 2021
This work is subject to copyright. All rights are solely and exclusively licensed by the Publisher, whether the whole or part of the material is concerned, specifically the rights of translation, reprinting, reuse of illustrations, recitation, broadcasting, reproduction on microfilms or in any other physical way, and transmission or information storage and retrieval, electronic adaptation, computer software, or by similar or dissimilar methodology now known or hereafter developed.
The use of general descriptive names, registered names, trademarks, service marks, etc. in this publication does not imply, even in the absence of a specific statement, that such names are exempt from the relevant protective laws and regulations and therefore free for general use.
The publisher, the authors and the editors are safe to assume that the advice and information in this book are believed to be true and accurate at the date of publication. Neither the publisher nor the authors or the editors give a warranty, expressed or implied, with respect to the material contained herein or for any errors or omissions that may have been made. The publisher remains neutral with regard to jurisdictional claims in published maps and institutional affiliations.

This Springer imprint is published by the registered company Springer Nature Singapore Pte Ltd.
The registered company address is: 152 Beach Road, #21-01/04 Gateway East, Singapore 189721, Singapore

Preface

The Internet of things (IoT) is the need of the hour, and it has escalated itself in every sphere of our day-to-day lifestyle starting from smart homes to smart agriculture, automobile sector to educational aids and automated workplaces to circumspect industries. According to 2016's Gartner's Hype Cycle for Internet of Things, IoT is currently at *peak of inflated expectations*. It is estimated that the number of IoT devices connected to the Internet will hike up to 34 billion by 2020. IoT is strengthened by the latest developments in wireless technologies, smart sensors, actuators, communication technologies, Internet protocols and cloud services. IoT is built on heterogeneous programming platforms and devices that lead to standardization issues. The management of huge amount of IoT data in a continuously expanding network gives rise to non-trivial concerns regarding data collection efficiency, data processing, analytics and security. With billions of IoT devices, which arise around us providing computing-intensive and delay-sensitive services, in many application scenarios, the low response latency for IoT services is achieved at the cost of computing-complexity of IoT devices and cloud. To improve efficiency and scalability of IoT applications, multiple computing paradigms emerge, such as mobile transparent computing, fog computing and mobile edge computing. Security-related aspects have been hindering a faster adoption of IoT devices. Aspects such as low processing power and small storage capacity of IoT devices contribute to their typically poor built-in security and forensics capabilities.

Wireless communication services in IoT have to compete with the existing users in radar, government and military communications, environmental monitoring and other IoT applications, which requires strategy to increase the efficiency of spectrum sharing among the enormous users in IoT. Therefore, it is essential to pursue fundamental research on new components, techniques and architectures to achieve energy-efficient sensing, communications and networking in a shared spectrum environment for IoT. Cyber-physical systems (CPSs) can provide both improved and new functionality with efficiency and convenience; but the increasing use of

CPSs and their application to key infrastructure components means that failures can result in disruption, damage and even loss of life. It is essential to develop CPSs which are reliable (or trustworthy).

Bhubaneswar, India
Santosh Kumar Pani
Manjusha Pandey

Contents

IOT: The Theoretical Fundamentals and Practical Applications 1
Santosh Kumar Pani, Siddharth Swarup Rautaray, and Manjusha Pandey

End-to-End Data Architecture Considerations for IoT 17
Alokika Dash

Domain-Specific IoT Applications 27
Sital Dash and Debashish Prusty

IoT in Rural Healthcare....................................... 37
Soumyajit Giri, Monideepa Roy, Sujoy Datta, and Animesh Goswami

IoT in Autism Detection in Its Early Stages 47
Sushama Rani Dutta, Monideepa Roy, Sujoy Datta, and Rupayan Datta

Significance of IoT in Education Domain 59
Hrudaya Kumar Tripathy, Sushruta Mishra, and Krushnakanta Dash

A Case for Unikernels in IoT: Enhancing Security
and Performance 85
Siddharth Choudhuri

Cloud of Things Assimilation with Cyber Physical System:
A Review 93
Yashwant Singh Patel, Manoj Kumar Mishra,
Bhabani Shankar Prasad Mishra, and Rajiv Misra

About the Editors

Dr. Santosh Kumar Pani is working as an Associate Professor at the School of Computer Engineering, KIIT Deemed to be University, Bhubaneswar, India. He received the Bachelor of Engineering in Computer Science & Engineering from the University College of Engineering, Burla, Master of Technology in from Utkal University and Ph.D. from KIIT (Deemed to be University). He has over eighteen years of teaching experience and served in many key administrative positions in the university. He has supervised several Ph.D. and postgraduate students in their Doctoral and Master thesis and published several research articles in reputed international journals and conferences. His research areas include program analysis, Internet of things, cloud computing and blockchain technology.

Dr. Manjusha Pandey is presently working as an associate professor in the School of Computer Engineering, Kalinga Institute of Industrial Technology Deemed to be University, Bhubaneswar, Odisha. She has completed her Ph.D. from Indian Institute of Information Technology, Allahabad. She has more than 80 research publications to her credit in journals and conferences of repute. Her research interest areas include, data analytics, machine learning, wireless sensor network, security and privacy in wireless sensor network, human–computer interaction and computer networks.

IOT: The Theoretical Fundamentals and Practical Applications

Santosh Kumar Pani, Siddharth Swarup Rautaray, and Manjusha Pandey

1 Introduction

In today's materialistic world, people are acing towards an immensely intelligent, highly productive, less expensive and time constrained lifestyle. **These needs exponentially increased the demand of automation**. Today, automation is spreading its roots very swiftly in all domains. Besides this, the continuous grooming of semiconductor industry has lead to a high growth in customized sensors which have eventually contributed to the field of automation. Parallely, IoT cloud industry is also accelerating at a fast pace with over 49 competitors [1] in the rapidly growing cloud vendor's market. According to 2016 Gartner's Hype Cycle for Internet of things [2], IoT is currently at "peak of inflated expectations". And according to facts stated by Cognizant [3], pertaining to the continuous growth in the IoT industry, the number of "things" will soon dwarf the number of people and mobile connections on the planet, accounting to a total of 7.4 billion mobile connections (including M2M). IoT is getting strengthened due to the latest developments in wireless technologies, such as smart sensors, actuators, communication technologies, Internet protocols and cloud services. Basically, IoT implements interconnection of different "things" which can be connected to cloud and monitored remotely through wired or wireless infrastructure. The creation of analytical models for predicting trends in IoT's data is automated by machine learning which enables algorithms to learn continuously with

---

S. K. Pani (✉) · S. S. Rautaray · M. Pandey
School of Computer Engineering, KIIT Deemed to be University, Bhubaneswar, India
e-mail: spanifcs@kiit.ac.in

S. S. Rautaray
e-mail: Siddharthfcs@kiit.ac.in

M. Pandey
e-mail: manjushafcs@kiit.ac.in

© The Author(s), under exclusive license to Springer Nature Singapore Pte Ltd. 2021
S. Kumar Pani and M. Pandey (eds.), *Internet of Things: Enabling Technologies, Security and Social Implications*, Services and Business Process Reengineering,
https://doi.org/10.1007/978-981-15-8621-7_1

the help of available IoT data and thus making the devices intelligent enough to take smart decisions [4]. Presently, machine learning is at peak of inflated expectations [2], i.e. it will take 2–5 years to reach plateau of productivity [5]. The management of huge amount of IoT data (i.e. data generated by sensors and other "things") will give rise to concerns regarding data collection efficiency, data processing, analytics and security. **By merging** M2M, SCADA, PLC, data analytics, cloud services and machine learning, IoT is bringing a huge transformation in the industry, i.e. from traditional embedded systems to modern IoT-based intelligent systems [6].

Internet of things (IoT):

Massive computations, security issues and mass storage, like limited communication capability, processing, energy and storage capacity, the need of cloud has become inefficient. These inefficiencies motivate us to combine the cloud with the IoT. As CC is the based technology, it consolidates various technologies and applications in order to gain maximum performance and capacity than the existing infrastructure.

Nowadays, the public are well known with the CC paradigm as well as the expansion of IoT. IoT was inducted by the Kevin Ashton in the year 1998 such that IoT is considered as the ubiquitous and Internet computing. The IoT technological revolution specifies the future network reachability and connectivity. In the term "Internet of things", the term, things, defines any objects on the earth, whether it can be any non-communicating or the communicating device. However, the objects specify the communicating nodes in the Internet in such a way that the data communication is progressed through the RFID tags. However, the device includes some smart objects such that the devices are the digital and the physical entity to achieve certain operations in network environment and humans. This is the reason that IoT is considered as not only the software and hardware paradigm but also it includes the social features and interaction as well. Accordingly, the architecture of IoT contains five different layers, namely perception, application layer, business layer, network layer and middleware layer [7]. Figure 1 represents the architecture of IoT layers.

Perception layer:

Perception layer is a layer situated at the lowest level in architecture diagram. Accordingly, the key purpose of perception layer is to receive the information from network environment. However, data sensing and the data collection process are achieved in the perception layer. Moreover, the entities, like RFID tags, camera, GPS, bar code labels and sensors, are situated at this layer. With the instance, the key role of this layer is to gather the data and to find the object or things in the network environment.

Network layer:

The second last layer is the second last layer, i.e. network layer which gathers the information from the perception layer at the bottom. The function of network layer is similar to that of the transport and network layer in the open system interconnection (OSI) model. However, the network layer is used to collect the information from the lowest layer and forward it to Internet. Accordingly, this layer only includes the gateway that has single interference that is linked to sensor network as well as

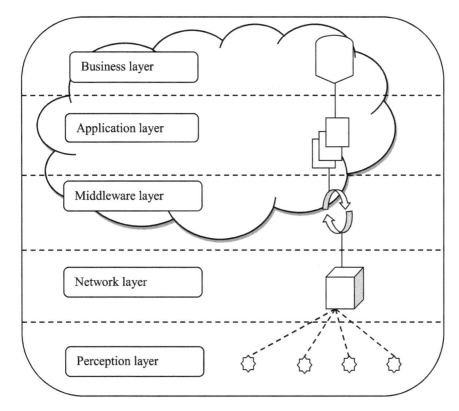

Fig. 1 Architecture of Internet of things layers

to the other networks. In certain scenario, the network layer includes information processing centre and network management centre.

Middleware layer:

The middle layer receives the information from the network layer. However, the major purpose of middleware layer is to store the data and to perform the service management process. Moreover, the middleware layer performs the processing of information system and automatically takes the decision based on the result. The middleware layer generates the output and passes it to next layer called application layer.

Application layer:

The application layer is used to perform the final arrangement of information. This layer receives the data from the previous middleware layer and offers the application of global management that present the data based on the data that is progressed by the middleware layer. However, the application layer defines the information in the appearance of smartness of home, smart vehicle tracking, smart health, smart farming, smart transformation and smart city.

Business layer:

In the business layer, the data that is received from the previous application layer and is moulded into some meaningful information, and then some other advance services will be generated from the existing data. Moreover, the data is processed into the knowledge form that makes the data to be more efficient and can use by the service providers to earn money. IoT operates on the basis of the machine-to-machine (M2M) communication such that M2M represents the communication that takes place among two different machines without any human intervention. Accordingly, in the IoT network, the non-connected units become the art of IoT through the data transmitting device, such as RFID tag and bar code that are eventually connected to Internet. However, the non-intelligent objects called things become the communicating nodes in the IoT terminology.

2 Applications

Iot has almost reached its plateau of productivity in sector of wireless sensor networks (WSN) while some other technologies based on IoT are still in their technology triggered phase. Overall, IoT has established itself appreciably into various domains. We will be exploring quite a few prominent domains below.

- IoT-based **Smart Buildings** are those devices where household devices or appliances can be monitored and controlled remotely using proper network architecture and standard protocols. Smart homes create comfortable living conditions by utilizing automation. Human comfort provided by these is categorized as thermal comfort, visual comfort, hygienic comfort, etc. [6]. The characteristics of a smart home are automation, multifunctionality, adaptability, interactivity and efficiency [7]. Automation in this field has reached to such an extent that we can even control staircase, washroom, corridor for efficient utilization of power and energy resources [8].
- **Smart Agriculture** monitors soil moisture content to trigger irrigation pumps, predicts type and fertility of soil to prescribe different fertilizers to the farmers. It can also predict the best suited crop based on the analysis of soil. Consequently, it enhances the crop quality as well as yield [9]. In addition to this, IoT is a boon to the hydroponic farming field [10].
- **Smart Health care** has the following use case scenarios.
  - Elder care: [11] IoT has the strength and solution for elderly living people. IoT can come up with assistance which is able to track and monitor patients. This plays a vital role in medication and has a mechanism to trigger in emergency situation.

- Physically handicapped: BlinDar [12], cap for visually impaired, smart blind guidance system's have been designed to improve the mobility and communication of both visually impaired people. It would help the blind person navigate alone safely. It works on the received signal, and the device would prompt a message and audio output to the user stating suitable actions.
- Smart alarm pillbox: This keeps a check on the patients' pill dosage. It informs the authorities when the pillbox gets empty and whether the patient has taken the pill or not and also wether it is right or not [13].
- IoT-based devices are also used to take preventive measures in ICUs [14] and OTs, e.g. pacemaker, blood monitoring devices, etc.
- Sleeping disorder: In this field, IoT is implemented to analyse the biosignals during sleep and automatically sleep disorders are detected [15].

- **Smart Environment** monitoring tools: Environmental monitoring tool is an IoT-enabled device which is used to detect any kind of unusual activities in the environment, i.e. for air, dump and water leading to alert system for common folk as well as municipal corporation. Collectively, this will result in healthy living and resource conservation. Smart UV and SPF monitoring system has also been developed for pollution control [16]. Also IoT is being used for prevention of border infiltration, an IoT-enabled system in which border infiltration by militants can be easily detected by the use of surveillance/reconnaissance drone embedded with thermal or infrared cameras as per the suitability of the operator.
- IoT-enabled water management system: It intelligently detects the turbidity, pH and optimum level of the water in its container. It can be used for household purposes or for municipal corporations [17]. Also monitoring of temperature, conductivity, relative humidity and gas concentration are measured using aquatic drone [18].
- Smart city IoT based—It includes smart garbage monitoring system which tracks the level of garbage bins [19] and the kinds of gas evolving out of the dustbins by providing a unique id to every trash can, so that signals can be generated to trigger immediate action, smart traffic control systems used for vehicular traffic prediction and control, etc. Also analytics are performed in real time in order to be used in different parts of the smart city to manage the oxygen level, smoke/hazardous gases and luminosity [20].
- IoT-based logistics and transportation industry's dynamic logistic markets require new products and services like RFID so that a logistic company can identify, track the parcels globally by the help of IoT. A Bluetooth module is used for communication which is short ranged [21].
- Industrial applications: IoT-based technologies are being implemented in almost all industries to enhance productivity by applying predictive maintenance and corrective measures to minimize faults. Piezoelectric energy harvesting-based access control system (P.H.A.S) sensor is motor that generates vibration, and it produces electricity through piezoelectric energy harvesting. Machine learning algorithm is used on vibration data to enhance the productivity [22].

- A recent research has proposed a big data analytics service based on IoT known as TSaaaS using time series data analytics to perform pattern mining on a large amount of collected sensor data [23].

We realize that a large variety of devices and sensor technologies with different functionalities from multiple manufacturers are required to implement IoT in the aforementioned fields, which creates an interoperability issue of interconnection and data exchange among this heterogeneous world of devices. Furthermore, this large pool of "things" makes it challenging to accomplish connectivity of these devices with the network and cloud. This is due to the cause that each device has different data format and different underlying protocol. Also different networking protocol and services are required to acquire communication among these variety of devices/things.

In order to solve this standardization issue, we need to promote interoperability between devices along with cloud and networking services. To effectively manage these devices, it is required to store their data and represent them in a standard desirable format, and at the same time we need to take care about the registration and compatibility of each device with cloud and network. Hence, we are proposing a common reference model that accommodates device side support, communication support, networking and cloud support along with data analytics and machine learning and domain-specific services. Not only this reference model provides support about modern devices but also it provides a platform to migrate traditional systems into it.

3 Associated Technologies and Industry Models

To develop a new IoT common standard architecture which can be used in order to develop any kind of IoT applications, we examined the traditional architecture, existing IoT models and this paves our way to build a model which encompasses the pros of existing model and sorts out their flaws introducing new required functionalities.

A traditional way of developing IoT applications is presented in Fig. 2, and this commonly used model consists of things/devices, network gateway, cloud platform and application server. Devices/"things" consist of multiple sensors, actuators, MCUs and other hardwares which fetch data from environment and send those data through communication/network gateway to the cloud servers. Communication gateway is required whenever things are not enabled with IP address (i.e. low-power devices with low computation which cannot run TCP/IP protocol suite). These "things" require a communication gateway which will coordinate different things and pass their information to the network gateway to send it over Internet. While, the high computation "things" like Raspberry Pi can directly send the data without any gateway [24] to the cloud. Internet gateway will mange all the traffic, and it may even act as firewall in some cases in order to protect the "things" and route those packets to their respective

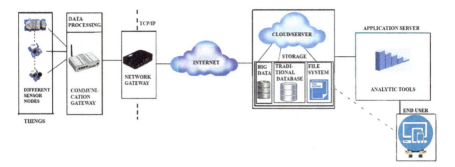

Fig. 2 Traditional M2M model

clouds and network servers. Further the cloud server is responsible for collecting, storing and analysing data subsequent to filtering out the unwanted data. These data can be later on stored, analysed and represented into desired format depending upon the user requirement.

We have studied traditional M2M model, and now we intend to explore existing Iot architectures of different organizations.

1. **Dell**:

They use an efficient application-specific gateway to secure data at the edge of the network and perform analytics [25] on the collected data to extract useful information and store them to the cloud servers. Bhoomi integration platform [26] for connecting cloud with the on-premises applications and data provides open-source cloud facilities [27].

This architecture emphasizes more on application like SCADA, home automation, vehicle automation, etc. Dell has innovated purpose-built edge gateways with specific I/O, form factor and environmental specifications to enable capture of data at the edge from a wide variety of sensors and equipment. The Dell edge gateway offers security and manageability tools along with unique features to help track and integrate data from even legacy network protocols, control systems and machines. In Dell's IoT architecture, different protocols are used for different applications, e.g. CAN bus protocol is suitable for vehicular automation. But there is no common program that is suitable for all applications. Provision for common gateway is missing, there is no device security, and data processing is scattered. However, the proposed architecture aims and has the capability to handle these problems efficiently by including the various **protocols** into the communication gateway. In Dell's architecture, customized programming codes can be written for different protocols to collect data, but once the data is sent to the cloud a common server side code handles the operations which is inefficient. Also, there is no dedicated device management unit. Data processing services are also not modifiable as per user's requirement.

2. **Azure**:

In the Azure architecture, the devices or things collect the data and send them to a cloud gateway. The cloud gateway makes the raw data available to the other back-end devices for processing. From here, the data is delivered to business applications or the users with the help of a user-friendly dashboard. This dashboard enables developers to have a clear view on their products. The end devices can connect to the cloud gateway directly or through the intermediate gateway. Then the Azure's most trusted cloud sends commands to specific devices and provides registration facilities, thereby it prevents the system from the vulnerable attacks by the alien devices. It also allows remote tracking and monitoring of the devices. The cloud stores the data to identify patterns and predict maintenance schedules. Azure IoT hub supports different features such that it can collect different data from different IoT devices like Raspberry Pi. Here we have to deploy Windows IoT core in Raspberry Pi [28] but for low-power devices we cannot install Iot core so by default of low-power devices are not interoperable with Iot core. We can develop, test and remotely monitor our IoT solution with device simulation. **User management, user registration and authentication are also there**. But user to device authentication is not available.

Microsoft Azure proposed an IoT architecture which has high scalability to hold millions of connections per hub. Azure IoT hub supports MQTT, AMQP as well as HTTPS but it does not directly support XMPP and COAP. Device-to-device(DDS) communication is also not directly provided here. No explicit categorization of devices is based on their resources as well as computational abilities.

3. **IBM IoT**:

IBM Watson IoT cloud platform [29] is an well-structured architecture framed to facilitate secure and ease at device connectivity, from stand-alone chips to smart appliances to complex industry solutions. It supports variety of devices along with different environment. The end devices send data to the cloud using HTTP or MQTT protocols. IBM cloud is supported by NOSQL database and time series databases. Device monitoring, diagnosis and management [30] are also possible in this. Identity as a service (IDaaS) is a new strong feature of this cloud. Real-time APIs allow detection of any changes in data and allow interaction between the systems or user at real time. Watson supports IBM analytics as well as geospatial analytics. The analytics unit analyses the collected data and sends control to the gateway so that only relevant data is sent to the cloud. Blockchain integration ensures security in online transactions. IBM Watson does not specify hardware abstraction layer or complex event processing. User management and user authentication service are there but it is not up to the mark. Different storage facilities are not available for different applications and types of data. Genius of things brought about a transformation combining Watson and AI.

4. **GCP**:

This GCP IoT architecture segregates the things into constrained devices (which cannot use TCP/IP) and standard devices. The constrained devices require to connect to the cloud through a gateway while the standard devices usually connect directly.

The ingest unit deals with decision-making, predictive maintenance and logging the data to show the health of various devices. The pub/sub [31] unit within the ingest handles any unnecessary changes of data due to spikes and prevents them from being reflected to the application. The data from here is sent in real time to the pipeline for further processing. Pipeline processes the data with the help of metadata, removes unnecessary data and provides desired data as the output. The data from the pipeline is sent to the analytical unit for further analysis. Finally, the data finds its way into the storage unit, where it may be stored in traditional databases or NoSQL databases. The stored or analysed data is kept available for the users to view it on the dashboard. IoT architecture as proposed by GCP has its own demerits such as only MQTT and HTTP protocols are supported, no interconnectivity among devices, provisions for remote control and management of devices is not available, Device authentication and registration are not up to the mark in this architecture, hardware layer abstraction is not included, and there is no edge analytics too as all the data is sent to the cloud directly for further processing. Users are not categorized, and user authentication and registration are not proper. Google provides the users with its state-of-the-art algorithms for machine learning used in search, as well as plenty APIs for such features as natural language processing, translation and computer vision which are remarkably good for fields like IoT.

5. **SmartThings**:

SmartThings is a smart home hub [32] designed and developed by Samsung. It works as a interpreter, suitable for a large number of devices, standards for different smart home systems where it is apt of making them work together. Also it allows remote monitoring of our "things". It is easy to install. SmartThings is built upon with Z-wave and Zigbee. It also supports IFTTT (If This Then That), i.e. trigger-based automation system. It works with third-party devices and has no limitation for number of sensors that can be added.

Hub does not support a native camera or pure protocols for connections and requires wired Ethernet connection. If the hub is in offline mode for a few seconds, whole system stops working, all apps and states become numb, and connection of hub to app is quite slow. Also the upgrading and integration from SmartThings hub are a very tedious task.

6. **ThingsWorx**:

ThingsWorx [33] is a popular data-oriented decision-making private cloud platform. It provides M2M and IoT-based IaaS. It provides tools and technologies that empower enterprises to rapidly develop and deploy powerful applications and augmented reality (AR) experiences. It is incorporated with search-based intelligence techniques(SEQUEAL). Its attractive features include runtime intelligent environment, zero coding facility, mobile interfaces mapped by APPs, event-driven execution engines at the servers, innovative 3D storage for millions of devices, data normalization, protocol translation, IoT device management, and device to cloud connectivity. ThingsWorx also supports Intel developed IoT hardware platforms. ThingsWorx's fallibility occurs in case of connecting huge number of devices as it supports attachment of only a limited number of devices. Also limited number of thing properties can be binded.

7. **AWS**:

AWS enables organizations to use the programming models, operating systems, databases and architectures with which they are already familiar. In addition, this flexibility helps organizations mix and match architectures in order to serve their diverse business needs. AWS delivers a scalable cloud computing platform that provides customers with end-to-end security and end-to-end privacy [34]. It is cost effective and scalable. AWS IoT core can process and route messages to AWS endpoints and to other devices reliably and securely. It allows us to easily connect devices to the cloud and to devices. AWS IoT core supports HTTP, WebSockets and MQTT. It provides authentication and end-to-end encryption throughout all points of connection. AWS provides an extensive set of services to handle analytics including data warehousing, business intelligence, batch processing, stream processing, machine learning and data workflow orchestration. It has a pre-configured Hadoop cluster along with Pig and Hive. It also supports Spark cluster. Also it has Amazon Recognition, Amazon Comprehend, etc., to provide cognitive services like video/image processing and natural language processing. But it has a few flaws like low-power devices, i.e. constraint devices cannot send data directly, and it does not support protocols like COAP, AMQP, XMPP, etc. Also AWS has a high cost and has a **high failure rate which requires us to spread our data geographically across several zone. Also EC2 has some serious performance and reliability issues as the load increases. AWS does not include enterprise grade support**.

4 Associated Technologies and Patents for IOT Applications

1. **Internet of things service architecture and method for realizing internet of things service (Patent No.: WO2011134318)**

This architecture of an Internet of things (IOT) includes multiple levels of IOT service platforms, wherein a superordinate IOT service platform is configured to manage one or more of the following function entities: IOT terminal, IOT terminal gateway, subordinate IOT service platform, special service platform and service gateway. The present invention also discloses a method for implementing an IOT service. The method includes the steps of the superordinate IOT service platform providing management for one or more of subordinate IOT service platform, special service platform and service gateway, wherein the management includes one or more of: registration, login, logout, data synchronization and heartbeat. With the present invention, the deployed industries or special service platforms can be integrated into a unified architecture, thus lightening the burden of the IOT service platforms. This architecture has been explicitly designed for M2M IoT applications and does not provide any categorization for devices/things. The architecture lacks provision of hybrid cloud, cloud monitoring and security, EDGE analytics, data validation and processing at either sides, i.e. service platform and things/devices. Hardware abstraction layer is completely missing, and there is no standardization for communication protocol to be used in different scenarios. Interfacing an intelligent device becomes a great challenge, and user cannot implant their customized business logic and rules. No care has been taken for big data, NoSQL model and storing of unstructured data.

2. **Integrated development tool for an Internet of things (IoT) system (Patent No.: US20170168777)**

The system contains an application comprising a GUI through which a developer is to specify a configuration for a new IoT device; a development database comprising configuration data related to different IoT device configurations, the IoT development application to utilize the data in the development database based on the configuration specified by the developer for the new IoT device; an IoT device engine to generate an IoT device profile responsive to the development application specifying input/output functions to be performed by the new IoT device; a client app engine to generate a user experience (UX) profile responsive to the development application specifying features of a client app or application related to operation of the new IoT device. This system does not make proper use of cloud service to store and analyse the data resulting to lack in complex data processing, analytics as per user demand, storing big data and unstructured data. This system also lacks in hardware abstraction layer and user abstraction layer, leaving many vulnerabilities open to infiltrate into the system.

3. **Adaptive and extensible universal schema for heterogeneous Internet of things (IOT) devices (Patent No.: US20140244833)**

The architecture is used to develop the apparatus for determining an association among Internet of things (IoT) devices, comprising a processor and a memory, wherein the memory comprises logic configured to receive, at a first IoT device, an identifier of a second IoT device; logic configured to obtain, by the first IoT device, a schema of the second IoT device based on the identifier of the second IoT device; and logic configured to determine, by the first IoT device, whether or not there is an association between the first IoT device and the second IoT device based on a schema of the first IoT device and the schema of the second IoT device, wherein the schema of the first IoT device comprises schema elements and corresponding values associated with attributes of the first IoT device and the schema of the second IoT device comprises schema elements and corresponding values associated with attributes of the second IoT device. This architecture lacks device \management, device registration and attestation unit. No provision for analytics by end users has been provided. Communication gateway is not standardized and not as per the latest trends. Whole communication is dependent on a access point, and once it fails whole system fails. EDGE analytics is not proper, and no common cloud architecture is used.

After going through an all-inclusive study, an abridged version of the aforementioned patents is provided in Table 1.

To analyse the previously achieved architectures for IoT, we went through some research papers and journals as Dhanalaxmi et al. [35] proposed to access the data parameters without intervention of main processor. In this method, a direct memory access controller (DMA) is used to fetch the real parameters. Cloud computing is not always preferable because huge amount of data is produced at the edge of the network, so sending of these data to the cloud servers without processing leads to

Table 1 All-inclusive details of recent patents filed for advanced IOT architectures

Serial No.	Patent no.	Description

1	WO2011134318	• Architecture of IoT includes multiple levels of IoT service platform
		• It manages a number of function entities
		  – IoT terminal, IoT terminal gateway, subordinate IoT service platform, special service platform, service Gateway
		• Its unified architecture lightens the burden of the IoT service platform
		• Explicitly designed for M2M IoT applications
		• Does not provide categorization for devices/things
		• Lacks provisions of
		  – Hybrid cloud, cloud monitoring, security, EDGE analytics, data validation, processing at either sides, i.e. service platform and things, hardware abstraction, standardization, storing of unstructured data
		• Interfacing an intelligent device is a great challenge
		• User cannot implement business logic and rules
		• No care taken for big data and NoSQL model

2	US20170168777	• Application comprising a GUI to specify configuration of new device • Development database comprising configuration data • IoT device engine to generate an IoT device profile • Client app engine to generate user experience profile • Lacks provisions of – Complex data processing, analytics as per user demand, nStoring big data and unstructured data, hardware abstraction layer, user abstraction layer • Does not make proper use of cloud service to store and analyse data • Vulnerability open to infiltrate into the system
3	US20142044833	• Determines association among IoT devices based on schema of first IoT device and schema of second IoT device • Lacks provision of – Device management, device registration, attestation unit, analytics, standardized communication gateway, proper EDGE analytics • No common cloud architecture exits • Single access point communication

more bandwidth consumption, storage issues, computation power and unnecessary overheads. To avoid this, we have to filter out unnecessary data at the device end. Micro data centres, cloudlet [30] and fog computing have been introduced for edge computing. Mobile phone integrated sensors can be enabled with IoT/WOT (Web of things) to enhance user's knowledge and perception to make better choices and take better decisions. Reconfigurable multi-sensor device called MoreSensor can be combined with a IoT management platform in the network domain to establish a IoT talk, i.e. MoreSensor device can "talk" to other IoT things. Low-end devices have resource constraints, e.g. limited memory, computation power and power supply. To build large-scale IoT systems, an appropriate OS to run on low-end IoT devices is required. None of the above studied architectures provide the standard platform to connect different types of things like constrained devices, standard devices and intelligent devices having the facility of EDGE computing within. There is no information regarding the selection of gateway under different types of network, communication channels/medium and protocols. No presence of a hybrid cloud that saves network bandwidth and reduces the chances of failure due to Internet connectivity problems. Absence of hardware abstraction layer allows the user to know from which device the data is being received, which is unnecessary. User abstraction layer is not included. There is no proper categorization of users and user authentication and management are missing. Adaptability of storage, to select a model according to different required situation (like using NoSQL model for unstructured data, ORDBMS model for structured data, Big Data model for huge data) is absent. No facility of rich APIs and libraries to ease IoT application development for developers and users. Deploying customized business logic with IoT cloud cannot be implemented easily. Preventive maintenance for the cloud and things is absent in various models to take care about devices and cloud system which reduces failure and assures zero down time. Device registration, attestation and monitoring are not up to the mark to meet upcoming future requirements. Data validation, data normalization along with complex data processing is completely absent which leads to garbled information and/or data inconsistency. Different analytics mechanisms in different hierarchies of clouds and devices termed as EDGE analytics (analytics within the device) are missing.

Even after studying almost all the big players for IoT cloud models, major IoT patents and significant IoT research papers/articles, we could not find a composite model that **accommodates** all the required functionalities such as

- It should support both constrained (low-power devices) and standard devices (smart devices).
- It should support heterogenous hardwares.
- It should support edge analytics and fog computing.
- This system should adapt older versions of hardwares having complex analogue circuit.
- It should support different programming environments to overcome inter-hardware communication challenges.
- It should support different data formats along with different server side technologies.

- It should have high scalability to adapt to the changing market.
- It should be smart enough to take preventive measures not only for the devices but also for itself.

5 Conclusion

This chapter aims to bring the overview of fundamental concepts of IOT and how its impact on the human day-to-day life has created a lot of ease and dependency leading towards more automation of processes. The varied applications of IOT as discussed in the chapter make the penetration of IOT more evident. Also we have tried to summarize the work done in the field implementations of IOT by detailing about different tools currently being used by industries for different applications of IOT. The effort has been made to include a section proposing the major requirements for further strengthening the currently used fundamentals and tools in IOT. The aim to include proposals for modifications is to enhance the applicability of the IOT hardwares as well as softwares platforms for enriched user experience, optimized performance and low cost. The interconnection of physical devices like PCs, smartphones, tablets, WiFi-enabled sensors, wearable devices and household appliances, etc., offered by IoT with the help of embedded sensors, actuators, programming logic and network connectivity is the strength that provides a platform for smoothe data exchange between devices. The evolution of IoT from the traditional embedded systems to M2M to current smart versions of Internet of things frameworks is detailed in the chapter to provide the researchers and practitioners in the domain area to have a consolidated information before moving towards new experimentations. As the penetration of IOT is increasing day by day, the requirement of standardization of the models used for interconnection among networks with heterogeneous hardwares and devices is what is the future of the IOT.

References

1. L.M. Dang, M. Piran, D. Han, K. Min, H. Moon, A survey on internet of things and cloud computing for healthcare. Electronics **8**(7), 768 (2019)
2. M.H. Miraz, M. Ali, P.S. Excell, R. Picking, A review on Internet of Things (IoT), Internet of Everything (IoE) and Internet of Nano Things (IoNT), in *2015 Internet Technologies and Applications (ITA)* (IEEE, 2015), pp. 219–224
3. J. Gogan, K. Conboy, J. Weiss, Dangerous Champions of IT Innovation, in Proceedings of the 53rd Hawaii International Conference on System Sciences (2020)
4. C.E. Ochoa, I.G. Capeluto, Strategic decision-making for intelligent buildings: comparative impact of passive design strategies and active features in a hot climate. Build. Environ. **43**(11), 1829–1839 (2008)
5. L.M. Camarinha-Matos, J. Goes, L. Gomes, J. Martins, Contributing to the Internet of Things, in Doctoral Conference on Computing, Electrical and Industrial Systems (Springer, Berlin, 2013), pp. 3–12

6. F. Tao, Y. Zuo, L. Da Xu, L. Zhang, IoT-based intelligent perception and access of manufacturing resource toward cloud manufacturing. IEEE Trans. Industr. Inf. **10**(2), 1547–1557 (2014)
7. C.L. Zhong, Z. Zhu, R.G. Huang, Study on the IOT architecture and gateway technology, in 2015 14th International Symposium on Distributed Computing and Applications for Business Engineering and Science (DCABES) (IEEE, 2015), pp. 196–199
8. H. Ghayvat, S. Mukhopadhyay, X. Gui, N. Suryadevara, WSN-and IOT-based smart homes and their extension to smart buildings. Sensors **15**(5), 10350–10379 (2015)
9. N. Gondchawar, R.S. Kawitkar, IoT based smart agriculture. Int. J. Adv. Res. Comput. Commun. Eng. **5**(6), 838–842 (2016)
10. N. Suma, S.R. Samson, S. Saranya, G. Shanmugapriya, R. Subhashri, IOT based smart agriculture monitoring system. Int. J. Recent Innov. Trends Comput. Commun. **5**(2), 177–181 (2017)
11. V. Vippalapalli, S. Ananthula, Internet of things (IoT) based smart health care system, in 2016 International Conference on Signal Processing, Communication, Power and Embedded System (SCOPES) (IEEE, 2016), pp. 1229–1233
12. Z. Saquib, V. Murari, S.N. Bhargav, BlinDar: An invisible eye for the blind people making life easy for the blind with Internet of Things (IoT), in 2017 2nd IEEE International Conference on Recent Trends in Electronics, Information & Communication Technology (RTEICT) (IEEE, 2017), pp. 71–75
13. A. Naditz, Medication compliance—helping patients through technology: modern "smart" pillboxes keep memory-short patients on their medical regimen. Telemed. e-Health **14**(9), 875–880 (2008)
14. P. Gope, T. Hwang, BSN-care: a secure IoT-based modern healthcare system using body sensor network. IEEE Sens. J. **16**(5), 1368–1376 (2015)
15. D.J. Cook, M. Schmitter-Edgecombe, Assessing the quality of activities in a smart environment. Methods Inf. Med. **48**(05), 480–485 (2009)
16. T. Robles, R. Alcarria, D.M. de Andrés, M.N. de la Cruz, R. Calero, S. Iglesias, M. Lopez, An IoT based reference architecture for smart water management processes. JoWUA **6**(1), 4–23 (2015)
17. S.M. Chaware, S. Dighe, A. Joshi, N. Bajare, R. Korke, Smart garbage monitoring system using Internet of Things (IoT). Ijireeice **5**(1), 74–77 (2017)
18. A.J. Trappey, C.V. Trappey, C.Y. Fan, A.P. Hsu, X.K. Li, I.J. Lee, IoT patent roadmap for smart logistic service provision in the context of Industry 4.0. J. Chinese Inst. Eng. **40**(7), 593–602 (2017)
19. M.R. Palattella, N. Accettura, L.A. Grieco, G. Boggia, M. Dohler, T. Engel, On optimal scheduling in duty-cycled industrial IoT applications using IEEE802. 15.4 e TSCH. IEEE Sens. J. **13**(10), 3655–3666 (2013)
20. M.R. Bashir, A.Q. Gill, Towards an IoT big data analytics framework: smart buildings systems, in 2016 IEEE 18th International Conference on High Performance Computing and Communications; IEEE 14th International Conference on Smart City; IEEE 2nd International Conference on Data Science and Systems (HPCC/SmartCity/DSS) (IEEE, 2016), pp. 1325–1332
21. S. Al-Sarawi, M. Anbar, K. Alieyan, M. Alzubaidi, Internet of Things (IoT) communication protocols, in 2017 8th International Conference on Information Technology (ICIT) (IEEE, 2017), pp. 685–690
22. E. Lefeuvre, A. Badel, A. Brenes, S. Seok, C.S. Yoo, Power and frequency bandwidth improvement of piezoelectric energy harvesting devices using phase-shifted synchronous electric charge extraction interface circuit. J. Intell. Mater. Syst. Struct. **28**(20), 2988–2995 (2017)
23. E. Ahmed, I. Yaqoob, I.A.T. Hashem, I. Khan, A.I.A. Ahmed, M. Imran, A.V. Vasilakos, The role of big data analytics in internet of things. Comput. Netw. **129**, 459–471 (2017)
24. A.N. Ansari, M. Sedky, N. Sharma, A. Tyagi, An Internet of things approach for motion detection using Raspberry Pi, in Proceedings of 2015 International Conference on Intelligent Computing and Internet of Things (IEEE, 2015), pp. 131–134
25. M. Satyanarayanan, P. Simoens, Y. Xiao, P. Pillai, Z. Chen, K. Ha et al., Edge analytics in the internet of things. IEEE Pervasive Comput. **14**(2), 24–31 (2015)

26. V.R. More, M.M. Bartere, Enterprise integration using Boomi tool. Int. J. Adv. Inf. Sci. Technol. **11**, 18–22 (2013)
27. D. Bruneo, S. Distefano, F. Longo, G. Merlino, A. Puliafito, I/Ocloud: adding an IoT dimension to cloud infrastructures. Computer **51**(1), 57–65 (2018)
28. B. Wilder, *Cloud Architecture Patterns: Using Microsoft Azure.* O'Reilly Media, Inc. (2012)
29. V. Scuotto, A. Ferraris, S. Bresciani, M. Al-Mashari, M. Del Giudice, Internet of things: applications and challenges in smart cities. A case study of IBM smart city projects. Bus. Process Manag. J. (2016)
30. M. Jia, J. Cao, W. Liang, Optimal cloudlet placement and user to cloudlet allocation in wireless metropolitan area networks. IEEE Trans. Cloud Comput. **5**(4), 725–737 (2015)
31. B. Grados-Licham, H.M. Bedón-Monzón, Software components of an IoT monitoring platform in Google cloud platform: a descriptive research and an architectural proposal (2020)
32. N. Bak, B.M. Chang, K. Choi, Smart Block: A Visual Programming Environment for SmartThings, in 2018 IEEE 42nd Annual Computer Software and Applications Conference (COMPSAC), vol. 2 (IEEE, 2018), pp. 32–37
33. Y. Yu, Quantitative Comparison of SensibleThings and ThingsWorx (2016)
34. W. Tärneberg, V. Chandrasekaran, M. Humphrey, Experiences creating a framework for smart traffic control using aws IOT, in 2016 IEEE/ACM 9th International Conference on Utility and Cloud Computing (UCC) (IEEE, 2016), pp. 63–69
35. B. Dhanalaxmi, G.A. Naidu, A survey on design and analysis of robust IoT architecture, in 2017 International Conference on Innovative Mechanisms for Industry Applications (ICIMIA) (IEEE, 2017), pp. 375–378
36. J.F. Valenzuela-Valdes, M.A. Lopez, P. Padilla, J.L. Padilla, J. Minguillon, Human neuroactivity for securing body area networks: application of brain-computer interfaces to people-centric Internet of Things. IEEE Commun. Mag. **55**(2), 62–67 (2017)

End-to-End Data Architecture Considerations for IoT

Alokika Dash

1 Introduction

Advances in shrinking form factor of embedded systems, ubiquity of networking, and low cost of sensors and processors has led to the proliferation of IoT (Internet Of Things) devices. An IoT system is often deployed to gather data and make decision about a system—at a fine granularity that is was not possible earlier. There has been significant effort towards the challenges of building and deploying IoT devices. Specifically, these challenges arise due to resource constraints that an environment may pose. For example, power consumption, form factor, cost per device, security, systems software (OS), and application software challenges need to be addressed before deploying an IoT. While the aforementioned are important, and there has been significant research work [1–5], an end-to-end view of the entire system in terms of data flow needs to be addressed as a first class architecture consideration *before* deployment. For example, in certain class of IoT devices, reacting in real-time to an anomaly in sensor value could be of utmost importance. In this case, *compute* and *latency* are the primary metrics of data i.e., the ability to compute and detect an anomaly and subsequently react. Further, this has to be done close to the IoT device (edge/fog) as opposed to the cloud or the datacenter due to strict latency requirements. In another class of IoT devices, analyzing months of data to learn about patterns could be of utmost importance. In this case, the *data storage format* is of utmost importance. In a third scenario, there could be a pipeline of systems that act on data where the first system in the pipeline acts on streams of data (i.e., near real-time) and the subsequent pipeline needs to run throughput oriented (batch, map-reduce) jobs. In all of the above cases, it is important to understand and architect the flow of data such that the SLA (service level agreements)—be it latency, throughput, and/or query processing rate, can be met. Therefore, an end-to-end data

A. Dash (✉)
Irvine, CA 92618, USA
e-mail: adash@alumni.uci.edu

oriented design consideration can lead to requirements that can be placed upon the architecture i.e., network considerations, processing power, RAM on each node etc. In the rest of the paper, we present the challenges and techniques to approach towards and "end-to-end" design of the system.

The rest of the paper is divided into two sections. In Sect. 2, a model for the end-to-end IoT system is defined and data considerations at each stage of this model is quantified. Such a model can help define the system specs at each stage. In Sect. 3, once the system is defined, design considerations for data are outlined that one must be aware of. Designing an IoT end-to-end system with such an approach can result in a predictable system and avoid problems due to either under or over engineering.

2 Modeling

Figure 1 shows a typical IoT architecture. The following are the various components of this architecture:

1. *IoT Device* This is the singleton unit of the system—a single senor/device that is capable of operating independently with the ability to detect (sense) and respond (transmit) to stimuli.
2. *IoT Deployment* A single IoT device in itself is not of much use. Typically, an entire group of IoT devices are deployed to sense and send their data to the next hop in the system.
3. *IoT Gateway* This is the first hop in the system that can gather data, perform limited computation on the gathered data, and the has the ability to interact with external servers. Examples of these systems could be anywhere from a local (on-prem) computer that gathers data to a cell phone tower servers. These systems are capable of limited compute on data such as finding aggregates, running inference models for AI, etc.
4. *Cloud/Datacenter* This is the last hop in an IoT architecture. The cloud and/or datacenter has the ability to provide virtually unlimited compute and data. The tradeoff being latency in response. In this paper, this is also referred to as "Backend system".

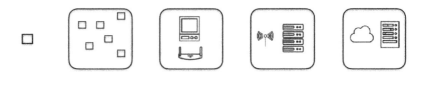

Fig. 1 IoT architecture overview

3 Data Formats

In this section, the data format at each step of IoT architecture shown in Fig. 1 is discussed. Specifically, a quantitative model is presented. While an exact quantitative model is very specific to an implementation, the goals of the data model presented in this section are to provide a first order approximation and bounds (upper limit) on data rate and format so that it can help guide the design of an actual (final) system.

3.1 IoT Device

The data format for IoT devices is typically in raw (binary) format. Given the power, cpu, and memory constraints, it is imperative that the raw sensor data is transmitted in the most efficient format. Therefore, transmitting in a format such as delta encoding [6] is an attractive solution. The reason for a delta encoded format are follows:

1. Most of the time the change in sensor value will be limited (i.e., within the low standard deviation). Given this assumption, the difference in subsequent sensor values can be encoded in few bits/bytes compared to the actual sensor value.
2. Delta encoding involves simple addition/subtraction (low cpu overhead) and maintaining minimal state in RAM (low memory overhead).
3. Delta encoding implies that the network overhead is minimal (low upload bandwidth) which is one of the main contributors to power.

If there are a total of N samples and the delta encoding is such that there are M intervals, and in each interval, there are m_i samples that sum up to N i.e., $\sum_{i=1}^{M} m_i = N$. The gain of using delta encoding is:

$$g = \frac{b \times N}{\left(b + \sum_{i=1}^{M-1} d_i \times m_i\right)} \quad (1)$$

where, b is the bit width for the sensor value. For example, $b = 8$ for 8-bit values, or $b = 64$ for unsigned long values. d is the number of bits used to encode the difference in subsequent values. For example, $d = 3$ if we know that the difference in sensor values is within 8 i.e., 2^3. Note that the delta encoding is efficient when $d \ll b$, and $N \gg M$ i.e., the changes in the sensor values are relatively small. These assumptions are not far from reality due to the fact that most of the times a cyber physical or IoT deployment works in a "normal" state where there are no anomalies that can cause a large delta between subsequent values.

3.2 IoT Deployment

An IoT deployment is a group of IoT devices that report and get response from a fixed set or a single IoT gateway. Since the IoT deployment is a group of IoT devices, the aggregate data that is generated due to the individual IoT device is of importance (i.e., the bandwidth or rate of data generation). If each IoT device, i, generates data every r_i^{th} second, and there are N IoT devices, then the edge gateway receives every j seconds, where: j is the GCD of the N IoT devices:

$$j = gcd(r_0, gcd(r_1, gcd(r_2, \ldots, gcd(r_{N-2}, r_{N-1})))) \quad (2)$$

Therefore, for an IoT gateway, the bandwidth the data processing requirements are as follows:

1. The IoT gateway must be able to receive data at a rate of every j seconds.

2. The amount of data received every j seconds is equal to:

$$\sum_{i=1}^{N}\left(D \mathrel{+}= \begin{cases} d_i, & \text{if } r_i \bmod = 0 \\ 0, & \text{otherwise} \end{cases}\right) \quad (3)$$

where, D is the amount of data (in bytes) received every jth second. Therefore, the data considerations will be to make sure that the network bandwidth on the edge has enough bandwidth to receive this data and the storage and buffering (RAM) has enough space to accommodate the accumulated data.

In summary, the data considerations put the following minimum requirements on the IoT deployment:

1. The ability to receive data at the peak rate without loss i.e., enough network bandwidth.
2. Enough buffering (RAM) to accommodate this data while it is being computed upon.
3. Enough storage space to persist this data if it is so desired.
4. Based on the data transformation requirements, enough network transmit bandwidth to push this data to the next level which is an IoT edge.

3.3 IoT Edge

The edge is the first hop in the flow of data which has dedicated compute resources for processing and making decision on data. However, the resources on an edge are still limited compared to a data-center or a cloud where there are virtually infinite compute and storage resources available. The flow of data between a gateway and the IoT edge can be modeled using queue theory.

One of the main reasons for having an IoT edge between a gateway and a server is that of reduced latency. The IoT edge can provide response to data in the order of high microseconds or few milliseconds due to the network proximity to the IoT gateway. Therefore, any algorithm or data-based decision made at the IoT edge will have to process streaming data in near real time. A few examples of such data-based decision algorithms are:

1. *Aggregates*: Statistical operations such as average, moving average, min-max etc. The streaming (incoming) data will have to be de-serialized to a standard format.
2. *Inference*: The de-serialized data can be passed into an neural-network inference model in order to classify and/or predict.

In both of the above cases, the data considerations are:

1. De-serialize data to a time-series data format.
2. Run stream-processing. For example using Apache flink, Spark, Kafka to name a few.
3. Optionally use other specialized processing such as Tensorflow if an inference is required.
4. The ability to—(i) optionally store this data on the server, and (ii) push the data to the backend server, and (iii) discard data based on a time-window if it is already stored.

In terms of data architecture considerations, the IoT edge plays the role of both a "backend" for IoT gateway and "frontend" for the backend-server (i.e., Cloud/Datacenter). The flow of data in this stage can be modeled using queueing theory, where the arrival rate is dictated by the IoT gateway and the departure rate is modeled based on the backend server. Figure 2 shows a queueing model that can be used to determine the system characteristics of an IoT edge.

Specifically, the following explain how a queueing model can be used to come up with a system specification for an IoT edge node.

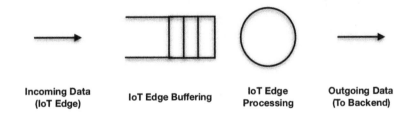

Fig. 2 IoT edge data queueing model

3.4 Incoming Data

This is the rate at which data arrives from one or more IoT gateways. This is driven by Eq. (3). The implications here are that the IoT edge must have sufficient network bandwidth to receive data at this rate. For example, if the IoT edge has a full duplex 1 GbE ethernet connection, then the receive rate for the IoT edge is capped at around 100 MB/s theoretically and practically around 80 MB/s given the IP layer overhead.

3.5 IoT Edge Buffering

Using Little's law $L = \lambda \times W$, one can determine the amount of RAM and processing required for the IoT edge node. In the IoT edge case, λ is the arrival rate of the incoming data. W is the rate at which data is processed which correlates to the computation (algorithm) required to process the data. Note that this computation can be in many different forms such as—(i) time required to run inference model on the data, (ii) time required to compute average and standard deviation, (iii) time required to transform the incoming data and write it to local storage. (Note that this involves matching CPU and IO bandwidth, which itself can be another queueing model). (iv) time required to transform data and push it to the backend server. (Note that this involves matching CPU and network upload bandwidth). Finally, L is the buffering required which in the IoT edge case is the amount of RAM (memory) available for application buffering. Note that this is just the memory required for application buffering and does not include memory requirements of the system itself such as operating system and other services (logging, daemons, ...) running on the IoT edge.

Since λ is fixed and known, one can perform design space exploration based on different values of RAM (i.e, L) and CPU (i.e., W). These can be based on power, cost, etc. Also, if the data needs to persist, this can further drive the requirements of the storage system i.e., choice of flash memory (SSD) and/or HDD. If the data storage requirements are known, Little's law can be applied further in order to calculate W i..e, $W = W_{cpu} + W_{storage}$ i.e., W is the sum of rates at which data is transformed by the (computation) and the rate at which data can be persisted.

3.6 Backend Servers

This is the last hop of the data flow. When designing an IoT system, one can consider the backend servers to be an on-prem data center or the cloud. The data storage here can be transformed from one format into another depending on the analytic needs. For example, the incoming data can be stored as a time series data base. This data for analytical purposes might need to be converted into a format that applications such as

Apache Spark, Cassandra, etc. can use to run analytical, map-reduce jobs. Therefore, the data may need transformation from time series data base into a NoSQL format. While not a full fledged ETL, the choice of data format is important.

4 Software Considerations for Data

Having established model for an end-to-end IoT system in the previous section, the table below summarizes the flow of data across each stage of an IoT system and describes the data considerations at each stage (Table 1).

Specifically, the considerations are as follows:

1. *Data Type* This is the layout of data. As we go from the IoT sensor towards the cloud, the nature of data turns from being bytes (raw/binary) to having a structure (Eg: SQL schema). As we move from left to right in Fig. 1, the data gathers meaning and form.
2. *Data Persistence* It is important to have a checklist at each stage on data persistence. The persistence further can be divided into short term and long term. Short term persistence implies less storage but needs guarantee that the short term storage has been processed before it can be overwritten by the next batch of data. For long term persistence, storage space is important and therefore, considerations such as compression becomes important which may not be necessary for short term persistence because the data is eventually ephemeral.
3. *Data Size* The data size along with persistence drives the combination of system RAM and storage and also helps in answering questions such as does one need in-memory database ? The data size in the cloud must consider questions as to how important the data is—does it need to be replicated, backed up, should one use RAID or not.
4. *Data Operations* The choice here is important because it drives what kind of software frameworks will run on the edge and on the cloud. Note that at each step, there can be multiple data processing and software frameworks and therefore it is best to prototype this ahead of time. For example, there are multiple software

Table 1 Data characteristics

Stage	Data type	Data persistence	Data size	Data operations
IoT device	Raw (binary)	Volatile	Bytes	Encoding
IoT deployment	Encoded	Volatile	KiBs to MiBs	Encoded, serialized
Edge	Streaming	Streaming and persistent	KiBs to MiBs	Aggregates, inference
Cloud (on-prem/public)	Schema and/or NoSQL	Non volatile	GiBs to TiBs	Map-reduce, query, AI

frameworks available for streaming—Spark streaming [7], Flink [8], Kafka [9], etc. Similarly, there are multiple framework available for machine learning—Tensorflow [10], Caffe [11], Pyspark [12] etc.

4.1 Data Interoperability

As data flows in an IoT system, the question of data interoperability i.e., as the output data from one software framework is passed to another software framework, does the data format needs a change ? If so, what is the cost of transforming data from one format to another. Specifically, what is the compute, memory, and storage cost of transforming data. This is especially an important issue given the wide matrix of software and open source frameworks available at each stage of data processing. One needs to answer the following questions:

(a) *Interoperability* How efficient it is to build a pipeline using two different frameworks. Questions such as—do they work together, is it required to write a lot of code for interoperability, etc.
(b) *Data Standard* Does the data need to be transformed between these software stacks or do they operate on a single standard data format.

5 Conclusion

In conclusion, when designing an IoT system, it is important to take a step back and enumerate the flow of data from sensor all the way to data at rest. Further, while enumerating, quantifying the rate and size of data flow can help answer design considerations such as minimum resource (CPU, memory, storage, and networking) required at every stage. Further, it is important to consider—(i) the data formats, and (ii) the data interoperability when making a choice for software framework used on data. Such a data and quantitative based approach to designing IoT can help guide the "end-to-end" design of an IoT system and alleviate hardware resource limitation issues that typically arise post deployment and are therefore, much harder to resolve.

References

1. S. Chen, H. Xu, D. Liu, B. Hu, H. Wang, A vision of IoT: applications, challenges, and opportunities with China perspective. IEEE Int. Things J. **1**(4), 349–359 (2014)
2. D. Blaauw, IoT design space challenges: Circuits and systems, *Symposium on VLSI Technology (VLSI-Technology): Digest of Technical Papers* (Honolulu, HI **2014**, 2014), pp. 1–2
3. J.B. Xu, Wendt, M. Potkonjak, Security of IoT systems: design challenges and opportunities, in *Proceedings of the 2014 IEEE/ACM International Conference on Computer-Aided Design (ICCAD '14)* (IEEE Press, Piscataway, NJ, USA, 2014), pp. 417–423

4. D.G. KorzunSergey, I. BalandinAndrei, V. Gurtov, Deployment of Smart Spaces in Internet of Things: Overview of the Design Challenges: Internet of Things, Smart Spaces, and Next Generation Networking pp. 48–59 (2013)
5. S.C. Mukhopadhyay, N.K. Suryadevara, Internet of things: challenges and opportunities, in *Internet of Things. Smart Sensors, Measurement and Instrumentation*, ed. by S. Mukhopadhyay, vol 9 (Springer, Cham 2014)
6. Delta Encoding. https://en.wikipedia.org/wiki/Delta_encoding
7. M. Zaharia et al., Apache Spark: a unified engine for big data processing. Commun. ACM CACM Homepage **59**(11), 56–65 (2016)
8. E. Friedman, K. Tzoumas, *Introduction to Apache Flink: Stream Processing for Real Time and Beyond* (O'Reilly Media, Inc. 2016)
9. Apache Kafka. https://kafka.apache.org/
10. TensorFlow. https://www.tensorflow.org
11. Caffe. http://caffe.berkeleyvision.org/
12. PySpark. http://spark.apache.org/docs/2.2.0/api/python/pyspark.html/

Domain-Specific IoT Applications

Sital Dash and Debashish Prusty

1 Introduction

Basically, Internet of Things (IoT) is a technology which generates network of things (devices, machine, etc.) which are capable to communicate with each other and share data through Internet [1]. There are lot of differences between the Internet of Things and the Internet. IoT can collect information about the connected objects, analyze it, and make decisions according to the analysis. IoT includes things having unique identities and is connected to the Internet [2, 3]. The Internet of Things (IoT) is the network of physical objects such as devices, instruments, vehicles, buildings, and other items immerse with electronics circuits, software, sensors, and network connectivity that allows these objects to collect and exchange data [4, 5]. The IoT allows objects to be sensed and controlled remotely [6, 7]. Generally, IoT is a network in which all physical objects are connected to the Internet through network devices or routers and exchange data [8, 9]. IoT is an intelligent technique which reduces human effort as well as allows easy access to physical devices [10, 11]. This technique also has autonomous control feature by which any device can control without any human interaction.

The objective of this chapter is to give idea about domain-specific application of IoT.

S. Dash (✉)
School of Computer Engineering, Kalinga Institute of Industrial Technology Deemed to be University, Bhubaneswar, India
e-mail: sitalfcs@kiit.ac.in

D. Prusty
Nokia Solutions & Networks, Espoo, Finland
e-mail: debashis.prusty@gmail.com

© The Author(s), under exclusive license to Springer Nature Singapore Pte Ltd. 2021
S. Kumar Pani and M. Pandey (eds.), *Internet of Things: Enabling Technologies, Security and Social Implications*, Services and Business Process Reengineering,
https://doi.org/10.1007/978-981-15-8621-7_3

2 Characteristics of Internet of Things (IoT)

The fundamental characteristics of the IoT are as follows [7, 11]:

- Interconnectivity: Using IoT, anything can be interconnected with the global network which gives information and they can communicate with each other using cloud-based infrastructure.
- Things-connected services: IoT-enabled networks connect things such as devices and machines which provides services such as security, privacy, and sharing between physical things and their associated virtual things.
- Heterogeneity: The devices in the IoT are heterogeneous as based on different hardware platforms and networks. They can communicate with other devices or service platforms in the same network or through different networks.
- Dynamic adaptation: The devices can adapt to new environment dynamically, and states of devices change dynamically.
- Vast scale: The number of devices that need to be managed and that communicate with each other will be at least an order of magnitude larger than the devices connected to the current Internet. Even more critical will be the management of the data generated and their interpretation for application purposes. This relates to semantics of data, as well as efficient data handling.
- Safety: As we gain benefits from the IoT, we must not forget about safety. As both the creators and recipients of the IoT, we must design for safety. This includes the safety of our personal data and the safety of our physical well-being. Securing the endpoints, the networks, and the data moving across all of it means creating a security paradigm that will scale.
- Connectivity: Connectivity allows network accessibility and compatibility. Accessibility is getting on a network while compatibility provides the common ability to consume and produce data.

3 Applications of IoT

The Internet of Things (IoT) applications span a wide range of domains including homes, cities, environment, energy system, retail, logistic, industry and health.

3.1 Home Automation

Home automation using IoT consists of several smart devices for different applications of lighting, security, entertainment, etc. All these devices are connected over a common network established by gateway. Home automation using IoT is explained as follows.

Domain-Specific IoT Applications

3.1.1 Smart Lighting

By using smart lighting for homes saves energy by adapting the lighting to the ambient conditions and switching on/off the lights whenever needed. Primary technology used for smart lighting uses solid-state lighting mainly LED lights and IP-enabled lights. Home achieves energy saving by sensing human body movements, generally when human walks toward home and also lighting condition of environment and controls light in smart lighting solution. Smart lighting system can be controlled remotely from IoT applications such as a mobile or Web application.

3.1.2 Smart Appliances

Modern people are using number of appliances such as TV, refrigerator, music system, and washing machine for their home. Managing and controlling these appliances can be difficult because each appliance has its own controls. Smart appliances make it easy to manage all these appliances remotely. For example, smart refrigerator has an RFID tag which stores number of items inside refrigerator and also send allot to the user when an item is low or out of stock.

3.1.3 Smoke/Gas Detector

Smoke detector is installed at home which is used to detect smoke in case of fire. Smoke detectors use optical detection or air sampling techniques which are capable to detect smoke. Smoke detectors generate allot in the form of signals to the fire alarm system. Gas detectors are used to detect harmful gases like carbon monoxide, liquid petroleum gag, etc.

3.1.4 Intrusion Detection

Home intrusion detection system consists of security cameras and sensors. Generally, it consists of door sensors to detect intrusions and raise alerts. Alerts can be in the form of SMS, email, an image or short video clip sent via email to the user (Fig. 1).

3.2 Cities

Cities use IoT devices such as connected sensors, lights, and meters to collect and analyze data. The cities then use this data to improve infrastructure, public utilities and services, and more. These cities are called as smart cities. Different smart city implementations are smart parking, smart lighting, smart roads, structural health monitoring, surveillance, emergency response, etc.

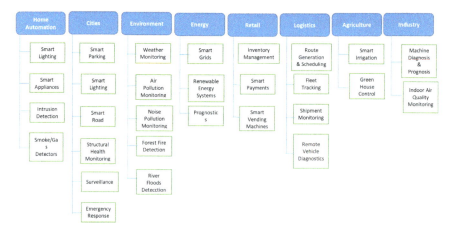

Fig. 1 Applications domains and relevant major scenarios

3.2.1 Smart Parking

In the crowded cities, it is very difficult to find a parking space. Smart parking system makes it easy for the driver to find a parking space, and this system is powered by IoT system which detects number of empty parking slots and sends information to smart parking application over the Internet. Smart parking application can be accessible by driver through smart phones, tablets, and in-car navigation systems.

3.2.2 Smart Roads

Sensors are deployed in the smart roads which can provide information on driving condition of the road, travel time estimation and also send alerts for poor driving conditions of the road, traffic congestion, and accident. This information sensed from the road is passed to the user via Internet to cloud-based applications.

3.2.3 Structural Health Monitoring

Structural health monitoring system is used to monitor the vibration levels in the structures such as bridges and buildings. This system uses a network of sensors to collect regarding vibration levels. These data are analyzed to assess the health of the structures and to detect crack and mechanical breakdowns.

3.2.4 Surveillance

For safety and security of cities, it is required to impose surveillance of infrastructure, public transport, and events which are organized in the cities. Surveillance can be achieved by a large connected network of security cameras. The video from the cameras can be aggregated in cloud-based scalable storage solutions.

3.2.5 Emergency Response

Another application of IoT system is to monitor critical infrastructure of cities such as buildings, gas and water pipelines, and public transport. IoT system also collects from large number of sensors and shares information for critical infrastructure using cloud-based architectures. Sensor data, audio, and video feed can also be analyzed in real time to detect adverse events using cloud-based architecture. Emergency responses are created and sent to the public in the form alert to re-routing of traffic, evacuation of the affected area, etc.

3.3 Environment

Environment monitoring such as weather monitoring, air pollution monitoring, noise pollution monitoring, forest fire detection, and river flood detection can be done by using IoT. Sensor can detect and measure any type of environment change.

3.3.1 Weather Monitoring

IoT-based weather monitoring systems are used to continuously monitor the climate condition and collect data from number sensors attached to the system such as temperature, humidity, and pressure. After collecting data, it sends the data to the cloud-based applications and back-end storage. The collected data are analyzed and visualized by cloud-based application, and weather alerts are sent to the subscribed user.

3.3.2 Air Pollution Monitoring

Gaseous and meteorological sensors are deployed by factories and automobile industry to monitor emission of harmful gases. This is called as IoT-based air pollution monitoring system. The data collected from sensor are analyzed to make decision on pollution control approaches.

3.3.3 Forest Fire Detection

Generally, IoT-based forest fire detection system can be made using wireless sensor networks. IoT-based forest fire detection system uses a number of sensors deployed at various locations of forest. Sensors collect the data related to temperature, humidity, light levels, etc., and analyze the data to generate alert for early fire detection.

3.4 Energy

Using the application of IoT, commercial and residential energy usage can be monitored and control. Some applications of IoT where energy can monitor and control are smart grids, renewable energy systems, and prognostics.

3.4.1 Smart Grids

Smart grid is an IoT-based system which is a data communication network integrated with electrical grids. In these networks, nodes are the sensors which collect and analyze the data captured in real time about power consumption, transmission, etc. Smart grid system gives information about utilization of power and also gives the solution to customer how to manage it.

3.4.2 Renewable Energy Systems

There are various renewable energy sources such as solar and wind which are available in nature which can affect grid stability and grid reliability and also give impact on power quality. IoT-based systems can be integrated with the transformer during interconnection to measure electrical variables and quantity of power into the grid which provides the solution to grid stability.

3.5 Logistics

IoT has a great impact on logistics and supply chain management. IoT makes easy to monitor shipments, to track the location of vehicle using fleet tracking system, to detect faults in vehicle using remote vehicle diagnostics, and to enroute shipment by choosing right route using route generation and scheduling application.

3.5.1 Route Generation and Scheduling

IoT-based system collects large amount of data from various sources and processes those data into information. IoT-enabled transportation system also collect huge amount of data from multiple sources to give services to the users. Route generation and scheduling system used the collected data to generate routes by combining transportation modes available and pattern of route and schedule the services according to the availability of vehicles.

3.5.2 Shipment Monitoring

Previously, once shipment is enroute, it is difficult to monitor it by owner. Nowadays, using IoT-enabled shipment monitoring system, it is easy to monitor the shipment. Shipment monitoring system uses sensors to monitor inside the containers and also send data to the cloud. Then, the collected data are analyzed to detect any damage of the product inside the shipment.

3.5.3 Fleet Tracking

Generally, fleet tracking system is used with vehicles. Vehicle fleet tracking system uses GPS technology to track location of the vehicle. Each vehicle is deployed with GPS technology. Fleet tracking system collects the data from GPS and generates allot in case of any scheduled route deviation.

3.6 Agriculture

World population always depends on agriculture. Nowadays, extreme changes in weather condition have large impact on agriculture industry. IoT technology can be used with agriculture industry to reduce waste and enhance productivity using smart irrigation and greenhouse control.

3.6.1 Smart Irrigation

Smart irrigation system uses soil moisture sensors which collect data about amount of moisture in the soil and then analyze the collected data and release water flow through pipes only when the water level of the soil goes blow the threshold level.

3.6.2 Greenhouse Control

IoT system plays a vital role in greenhouse control. Greenhouse is nothing but a glass structure with a plastic roof that provides required environment for plants to grow. The inside environment of greenhouse can be controlled and monitored to provide proper condition for plant growth. In IoT-enabled greenhouse, there are sensors inside the greenhouse which collect data and store in cloud where analysis data will be carried out for optimization of controlling strategy.

3.7 Retail

Nowadays, IoT applications are more popular in retail industry. Retail industry uses smart payment system to reduce cash transaction, smart vending machines to reduce human intervention to give products to the customers, and inventory management system to check their stocks.

3.7.1 Inventory Management

Inventory management was a tedious task previously. If someone is failed to properly manage inventory, then there may be over stock of products which requires huge storage or may be some products are out of stock, which leads to loss revenue. IoT-enabled inventory managements make it easy to check stock and keep the track of stocked product. RFID tags are attached with the products. By using these tags, products can be tracked in real time.

3.7.2 Smart Payments

Nowadays, everyone is talking about contact-less payments. This can be achieved by smart payment systems. Contact-less payment is powered Near-Field Communication (NFC) and Bluetooth. Users of smart payment system can store their credit and debit card information in NFC-enabled smart phones and make payments by taking the mobile phone to the sale terminal.

3.7.3 Smart Vending Machine

Smart vending machine requires Internet connection and allows to monitor inventory levels, pricing of products, and also contact-less payments. Smart payment system is also required for smart vending machines. Sensors are deployed with smart vending machines which monitor operations of vending machine, collect, and send data to the cloud which is analyzed for maintenance of the machine [2, 12].

There are also many more applications of IoT. Another two domains are there such as industry and health and lifestyle where IoT is used vastly. In any industry, IoT is used to diagnosis and prognosis of machines, to monitor air quality and also smart lighting system. There are lots of application of IoT in health care and lifestyle. Various wearable devices are available which is used to monitor health and fitness of user. Many people are wearing smart watches, smart glasses, wristband, and neckband to take advantages of IoT.

4 Conclusion

Internet of Things is a new era of the Internet. Internet has changed the lifestyle of people drastically. Day by day, virtual connectivity between people is increasing instead of physical connectivity. IoT gives the platform for this virtual connectivity with vast ranges of application. By using IoT, anyone can do anything with any media at anywhere. We observe that IoT will be acquired the overall Internet in near future. Since IoT-enabled devices are easy to use and easy to maintain, people are attracting toward this.

References

1. O. Vermesan, P. Friess, P. Guillemin, S. Gusmeroli, et al., in *Internet of Things-Global Technological and Societal Trends*. Internet of Things Strategic Research Agenda (Chapter 2, River Publishers) (2011)
2. M. Serrano, P. Barnaghi, F. Carrez, P. Cousin, O. Vermesan, P. Friess, *Internet of Things Semantic Interoperability: Research Challenges, Best Practices, Recommendations and Next Steps*. European research cluster on the internet of things, IERC (2015)
3. ITU-T, Internet of Things Global Standards Initiative, http://www.itu.int/en/ITU-T/gsi/iot/Pages/default.aspx
4. IoT: https://dzone.com/articles/the-internet-of-thingsgateways-and-next-generation
5. K. Rose, S. Eldridge, L. Chapin. *The Internet of Things: An Overview Understanding the Issues and Challenges of a More Connected World*. The Internet Society (ISOC) (2015)
6. Dr. Ovidiu Vermesan SINTEF, Norway, Dr. Peter FriessEU, Belgium, "Internet of Things: Converging Technologies for Smart Environments and Integrated Ecosystems", river publishers' series in communications (2013)
7. Dr. Ovidiu Vermesan SINTEF, Norway, Dr. Peter FriessEU, Belgium, "Internet of Things–From Research and Innovation to Market Deployment", river publishers' series in communications (2014)
8. H. van der Veer, A. Wiles, *Achiveing Technical Interoperability—The ETSI Approach*, 3rd edn. ETSI White Paper No. 3, http://www.etsi.org/images/files/ETSIWhitePapers/IOP%20whitepaper%20Edition%203%20final.pdf (2008)
9. https://www.ida.gov.sg/~/media/Files/Infocomm%20Landscape/Technology/TechnologyRoadmap/InternetOfThings.pdf

10. Martin Serrano, Insight Centre for Data Analytics, Ireland, Omar Elloumi, Alcatel Lucent, France, Paul Murdock, Landis + Gyr, Switzerland, "ALLIANCE FOR INTERNET OF THINGS INNOVATION, Semantic Interoperability", Release 2.0, AIOTI WG03—IoT Standardisation (2015)
11. http://www.reloade.com/blog/2013/12/6characteristicswithin-internet-things-iot.php
12. http://tblocks.com/internet-of-things

IoT in Rural Healthcare

Soumyajit Giri, Monideepa Roy, Sujoy Datta, and Animesh Goswami

1 Introduction

The Internet of things (IoT) is the implementation of interconnectivity between physical objects and electronic devices [1, 2]. The availability of Internet connectivity in such devices and sensors helps these devices to communicate with others over the Internet and enables us to control and monitor various systems from remote locations.

One such application of IoT is that it can be used in tracking the spread of infectious diseases in remote and populated villages through the help of such interconnected sensors and devices.

Tracing the source of an epidemic and subsequently generating timely alerts is very important, if we want to effectively control and contain an epidemic [3, 4]. Controlling the epidemic becomes even more difficult if it is in rural areas, as the rural areas are already constrained by poor healthcare infrastructure and network connectivity.

So to effectively contain the outbreak of any contagious disease, and to minimize its spread, we need to have a very fast and efficient alert system in place.

Epidemics can spread swiftly through a population, and their effects can persist over long durations. If not detected and dealt with at a preliminary stage, the toll

---

S. Giri · M. Roy (✉) · S. Datta
KIIT, Bhubaneswar, India
e-mail: monideepafcs@kiit.ac.in

S. Giri
e-mail: soumyagirigo@gmail.com

S. Datta
e-mail: sdattafcs@kiit.ac.in

A. Goswami
Confiance Mobility, Kolkata, India

© The Author(s), under exclusive license to Springer Nature Singapore Pte Ltd. 2021
S. Kumar Pani and M. Pandey (eds.), *Internet of Things: Enabling Technologies, Security and Social Implications*, Services and Business Process Reengineering,
https://doi.org/10.1007/978-981-15-8621-7_4

from such epidemics can rise exponentially and in extreme cases can wipe out large parts of a population.

People in rural areas are more prone to such attacks because of various reasons like lack of awareness, poor connectivity, lack of proper sanitation and cleanliness, and comparatively less access to adequate medical facilities. Therefore, if we deploy an epidemic detection and management system in such areas, it can help to effectively contain and minimize the spreading of a particular epidemic.

In our proposed framework, a particular region consisting of a number of villages is further divided into smaller areas and each of which has a number of medical kiosks run by healthcare workers. Each kiosk is equipped with various sensors and other electronic medical devices for measuring blood pressure, body temperature, heartbeats and various other physical parameters. The body parameters of the villagers are measured through these devices and the results are accumulated and monitored over the network [5–7]. An epidemic alert is generated in a particular region if the number of villagers who report the symptoms of the designated disease crosses a pre-decided threshold value, which is calculated as per the cumulative patient information of an area, and reaches the concerned health department on a real-time basis. It also displays the number of infected, semi infected and uninfected patients on a current time basis.

2 The Current Scenario in Rural Areas

People in rural areas suffer from several air- or water-borne diseases, which are infectious or contagious, such as typhoid, diarrhoea, malaria, infectious hepatitis, tuberculosis, worm infestations, measles, whooping cough, respiratory infections and pneumonia. Any of these might cause an outbreak of an epidemic among the population of a particular area.

People in villages also have to suffer from intermittent transport and network linkages, and little or non-existent medical support. Ways in which IoT can be of great help are: continuous monitoring of villages, and for raising awareness of epidemic outbreaks or other contagious diseases in rural and remote areas with the corresponding authorities.

Here, the real-time health data of the patients is collected from the kiosks to gauge the extent of the epidemic development. The vital signs from the wearable devices from the kiosk are received by the system, along with the pathological test reports of the patients, and sent to the central health base station. If the count of the infected persons in a specific area exceeds a particular threshold value, then an alert is sent to the concerned authorities [8, 9].

3 Objectives of the System

The main objectives of the automatic alert generation system are:

- Collecting the patient's vitals from wearable sensors and other electronic devices present in the kiosk in real time.
- Sending epidemic outbreak information based on the consolidated information of an area if the count of infected persons from a specific area increases beyond a threshold value.
- Providing infected, semi-infected or safe patient details based on their current health signs.

4 The Proposed Architecture

The Remote Area Epidemic Detection System (RAEDS) framework proposed by us is described as follows:

(a) **Healthcare Kiosk**:
(b) **Gateway Node**:
(c) **Central Healthcare Repository**:
(d) **E-Health Centre**:

The concerned region is divided into smaller areas, which are again divided into smaller areas, each containing some kiosks as shown in Fig. 1.

The framework consists of the following parts:

(a) **Healthcare Kiosk**: Here, there are kiosks, each of which connects to its own gateway for a particular area. The kiosks are setup as per their geographical locations. The longitude and latitude from the GPS are used to track a kiosk. Villagers visit their nearest kiosk. The sensors at the kiosks are used to monitor the symptoms of a patient. The readings of each patient are stored along with a patient ID for later access and notifications if necessary.
(b) **Gateway Node**: This node receives all the data from the kooks within a particular area and sends it to the Central Healthcare Repository.
(c) **Central Healthcare Repository**: The alert is generated here as per the proposed algorithm. The central repository holds the reference list for the symptoms, $E = \{S_1, S_2, S_3..., S_n\}$ where E is the epidemic situation and is the combination of symptoms like $S_1, S_2, S_3..., S_n$. When a kiosk sends a patient's current symptom details to the server, all the symptom details of the patient are stored in the patient repository, along with the patient ID as shown in Table 1. To verify the infection status of a patient, the list of current symptoms of a person from a specific region is matched with the reference list of epidemic symptoms in the central repository. If the lists match, then that particular individual is infected and is denoted as red circles. If the lists match partially, then he/she is partially/semi-affected and denoted as blue circles. If they do not match, then the person is

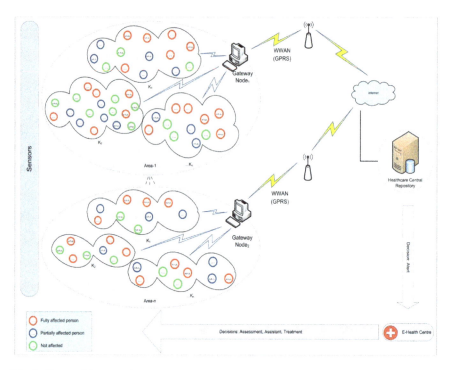

Fig. 1 Proposed framework

Table 1 Patient symptom list measured by the kiosks

Area.Kiosk.Patient	P_1	P_2	P_n
$A_m.K_1.P_n$	S_1, S_4, S_5, S_7	S_2, S_4	S_2, S_7, S_n
$A_m.K_2.P_n$	S_4, S_7	S_1, S_5	S_2, S_3, S_4, S_n
$A_m.K_3.P_n$	S_2, S_3, S_4	S_3, S_9	S_5, S_n
$A_m.K_c.P_n$	S_1, S_n	S_1, S_2, S_3	S_3, S_n

not infected and denoted as a green circle. When the count for fully affected persons crosses the threshold value for a particular area, an alert is sent to the E-health centre.

(d) **E-Health Centre**: Based on the alerts received by the centralized repository, the E-health centre accordingly decides and provides the required healthcare assistance to the areas which have been affected. They also track fully and partially infected patients for sending the requisite medical treatment instructions to him/her if needed.

5 Implementation

The patients in the villages are sent to visit their respective kiosks for checkups. The kiosks are handled by groups of health workers who monitor the patients for the symptoms with the necessary equipment. The cumulative results of these records are sent to a central repository and are used to determine whether a particular area is under threat of an epidemic or not [10].

The algorithm that is applied to the records for generating the alert is as follows:

Algorithm 1: Algorithm for finding epidemic Situation

```
1  FindEpidemic ()
       Input : patientSymptomList[],
               epidemicSymptopmList[]
       Output: epidemicAlert, area.Epidemic.Patient
2      foreach P[i] in A[i] do
3          ▷ For all Patient[i]:p_1..p_n in an Area A[]=a_1..a_m if
           (area.patientSymptomList[] equals
           epidemicSymptopmList[]) then
4              area.kiosk.patientFlAft.epidemic[]++;
5                                  ▷ Fully affected patient
6              area.epidemic[]++; ▷ Epidemic no. in an Area
7          else if (area.patientSymptomList[] contains
           epidemicSymptopmList[]) then
8              area.kiosk.patientPrtAft.epidemic[]++;
9                                  ▷ Partially affected patient
10         else
11      |      continue     ▷ Continue to next iteration

12     foreach epidemic in area.epidemic do
13         if (area.epidemic.count >(Threshold)) then
14             alert.epidemic[]++;
15             alertHealthCare();   ▷ Alert to Healthcare
16         else
17      |      continue

18     resultmap.add(alert.epidemic[]);
19     resultmap.add(area.kiosk.patientFlAft.epidemic[]);
20     resultmap.add(area.kiosk.patientPrtAft.epidemic[]);
21     return resultmap[];
22                          ▷ Return Alert and Patient details
```

While the checkup is done, the algorithm makes a comparison of the list of patient's symptoms with the list of epidemic symptoms. If the lists match fully, then the patient is marked as fully affected, and the counter for the fully affected patient as well as the counter for epidemic alert for that area is incremented by one for that particular kiosk centre. If the lists match partially, then the patient is marked as partially/semi-affected and the counter for partially affected patients as well as the counter for that particular kiosk centre is incremented by one. However, in this case, the area epidemic count is not incremented as the patient is not counted as under fully infected. While doing the checkups, the counter for epidemic is constantly monitored to see if it exceeds the threshold for the epidemic limit. The value of the threshold is determined by the experts based on the density of the population of a particular area. In case the area epidemic count exceeds the threshold value, the central repository will raise an alarm to the healthcare centre. Finally, the complete details of the fully affected patients, because of which the alert was raised, as well as those of the partially affected patient are sent. Adequate actions as deemed necessary are then taken, depending on the situation.

6 Deliverables

The testing for the entire process was done on Android devices running the client UI and the server [11, 12]. Figure 2 is a screenshot where the epidemic alert is raised. In this case, we have taken malaria as the test case in focus. As per medical documentation, the symptoms of malaria include: fever, nausea, vomiting, headache, malaise, diarrhoea, sweating and chills, abdominal pain, body aches, sore throat, running nose, difficulty breathing, cough and convulsions. When the patient symptoms are

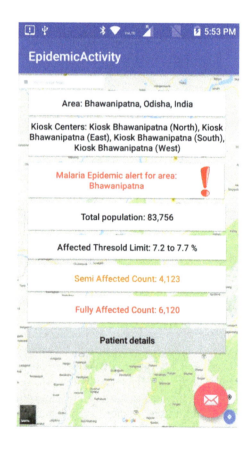

Fig. 2 Screenshot displaying the epidemic alert

IoT in Rural Healthcare

sent from the client application to the server, they are stored area-wise in a JSON-based storage system. As soon as a new entry from a new patient comes into the JSON storage, it will store it along with the ID in the [area.kiosk.patient] format, to be retrieved for later use if needed. In case a patient exhibits all the symptoms from the symptom set, then the person is marked as a fully affected person. If the patient exhibits some of the symptoms, then he is marked as semi-affected.

As per Center for Disease Control and Prevention (CDC), the threshold limit for this app has been set as 7.2–7.7% of the cumulative population of the region. When the counter for the fully affected persons exceeds the threshold limit, an alert is raised for that area. As shown in the screenshot, a selected ahead is divided into four parts: East, West, North and South. Kiosks are deployed in all of the sectors. The patient details section contains details regarding fully and semi-affected population who may need quarantine.

Figure 2 is a screenshot displaying the epidemic alert for malaria. Figure 3 is a screenshot showing the area in which there is no epidemic alert. So the counter for fully affected persons in these areas is under the threshold limit. The concerned authorities can arrange for subsequent action after studying this on a real-time basis.

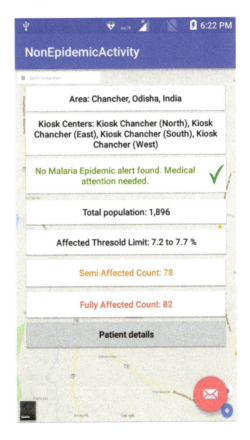

Fig. 3 Screenshot of alerts for malaria epidemic in a specific area

If there are two areas with the same population density and death and recovery rates, but differ with respect to the number of affected persons, then the region with the higher rate of affected people will be flagged as the region with higher epidemic rate.

Two factors, given below, can help in detection of the epidemic outbreaks in a given area, taking the two aforementioned scenarios in account.

1. The time is divided into several intervals and denoted by θ. During each of these intervals, the number of total infected people is counted. In a unit area, a patient is considered infected if the person is affected by the symptoms (S) for θ intervals at time **t** and denoted by S, t. If the person is currently uninfected, it may be possible that the person will exhibit more symptoms at a later stage.
2. The total number of individuals, who are affected in a unit area, is based on the number of total outbreaks in a kiosk from that area. To identify the population who has been infected for intervals for time t, we denote them as $P\,t$,. The total number of infected persons from the kiosks of an area is defined E_p which is the summation of the individuals who are diagnosed in the intervals within an area. Moreover, the population that is unaffected may also be more susceptible to it during the $t - 1$ or t_o time. Also, we use P_t symbol to denote the number of patients who have actually suffered from the infection in the duration of $(t - 1)$ to t interval.

Figure 4 shows the schema designed for the entire epidemic situation. Here, the "Affected Individuals" denotes how many people are affected by the symptoms currently and they are suffering from in the intervals. The arrows represent the patients requiring medical attention until he/she has completely recovered during the intervals. Finally, the "total number of ill people" is denoted by En for which the epidemic alert is generated.

Fig. 4 Control of epidemic

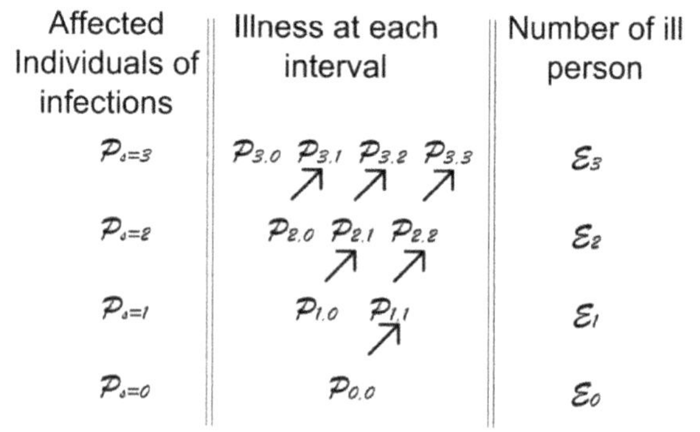

Affected Individuals of infections	Illness at each interval	Number of ill person
$P_{s=3}$	$P_{3.0}$ $P_{3.1}$ $P_{3.2}$ $P_{3.3}$ ↗ ↗ ↗	ε_3
$P_{s=2}$	$P_{2.0}$ $P_{2.1}$ $P_{2.2}$ ↗ ↗	ε_2
$P_{s=1}$	$P_{1.0}$ $P_{1.1}$ ↗	ε_1
$P_{s=0}$	$P_{0.0}$	ε_0

7 Conclusion

In this chapter, we have proposed a framework which will generate an alert if the affected population in a region exceeds a threshold limit. We have designed the architecture based on WBANs and sensors for collecting the real-time body vital signs of patients. If implemented on a large scale, this application has the potential to significantly help in detecting and containing epidemics in affected areas.

References

1. V. Chan et al., Mobile e-Health monitoring: an agent based approach. IET Commun. **2**(2), 223–230 (2008)
2. Boric-Lubecke et al., E-healthcare: Remote monitoring, privacy, and security, in IEEE MTT-S International on Microwave Symposium (IMS), 1–6 June 2014, pp. 1–3
3. Z. Zhang et al., Epidemic control based on fused body sensed and social network information. 32nd International Conference on Distributed Computing Systems Workshops (ICDCSW) (IEEE, 2012)
4. D. Malan et al., Codeblue: an ad hoc sensor network infrastructure for emergency medical care. International Workshop on Wearable and Implantable Body Sensor Networks, vol. 5 (2004)
5. B.P.L. Lo et al., Body sensor network wireless sensor platform for pervasive healthcare monitoring (2005), pp. 77–80
6. C. Otto et al., System architecture of a wireless body area sensor network for ubiquitous health monitoring. J. Mob. Multimedia **1.4**, 307–326 (2006)
7. O. Chipara et al., Reliable clinical monitoring using wireless sensor networks: experiences in a step-down hospital unit. Proceedings of the 8th ACM Conference on Embedded Networked Sensor Systems. ACM (2010)
8. L.A. Meyers et al., Network theory and SARS: predicting outbreak diversity. J. Theor. Biol. **232.1**:71–81 (2005)
9. S. Sudarshan et al., AutoHS: the intelligent hospital search. Int. Conf. Intell. Networking Collaborative Syst. (IEEE, 2014)
10. I. Mohomed et al., Harmoni: context-aware filtering of sensor data for continuous remote health monitoring. Sixth Annual IEEE International Conference on Pervasive Computing and Communications (IEEE, 2008)
11. N.R. Prakash et al., Body area network based health monitoring of critical patients: a brief review. Int. J. Instrument. Control Syst. **2** (2012)
12. S. Amit et al., Dynamic connectivity establishment and cooperative scheduling for QoS aware wirelessbody area networks. IEEE Trans. Mobile Comput. **117**, 2775–2788 (2018)

IoT in Autism Detection in Its Early Stages

Sushama Rani Dutta, Monideepa Roy, Sujoy Datta, and Rupayan Datta

1 Introduction

The internet of things (IoT) is an interrelation of many things like animals, plants, objects, people, machines, computers, many digital devices which are having their own unique identifiers (UIDs). These devices communicate with each other through the Internet with their UIDs. IoT has a great impact on cloud computing, research, data processing, in wireless communication etc. The cloud computing provides different types of OS, software, storage devices and data processing facilities. The Internet of Things (IoT), has a great impact in the health domain. Some of the advantages of IoT in healthcare are: helps to lower the expenses of the patients, provide better treatment, better disease control, lower the mistakes in diagnosis, and provide better care for the patients.

We have incorporated IoT in the existing autism diagnosis process to help the doctors to arrive at faster and more accurate diagnoses. Our proposed autism diagnosis system used the Electronic Health Record (EHR). The EHRs are the database of patient's reports which can be accessed from the hospital by the authenticated users from any place, and is possible through the help of internet services. The patient data

---

S. R. Dutta · M. Roy (✉) · S. Datta
School of Computer Engineering, KIIT University, Bhubaneswar, Odisha, India
e-mail: monideepafcs@kiit.ac.in

S. R. Dutta
e-mail: sushamadutta@gmail.com

S. Datta
e-mail: sdattafcs@kiit.ac.in

R. Datta
Ericsson, Kolkata, India
e-mail: rupayan.datta@ericsson.com

© The Author(s), under exclusive license to Springer Nature Singapore Pte Ltd. 2021
S. Kumar Pani and M. Pandey (eds.), *Internet of Things: Enabling Technologies, Security and Social Implications*, Services and Business Process Reengineering,
https://doi.org/10.1007/978-981-15-8621-7_5

are collected through a set of interconnected devices from various regions and stored in the EHR. The data stored in the EHR available in the cloud can be processed according our requirements. Our proposed system categorized the type of autism patients in different categories in the EHR, which will help us to identify the disease easily. These pre-processing of data can be done in cloud environment, which makes our diagnosis system faster.

An Electronic Health Record (EHR) is used to store the patient details like the demographic data, symptom details with disease identified etc., and are available in the various health departments. The EHR is generally updated periodically by the health department and uploaded to the cloud storage and can be accessed by the authenticated persons from any place for different uses. The updated EHR can be accessed from any rural area through internet for the processing of the autism diagnosis system. Health related reports can be accessed from the hospital by the authenticated users from any place through the help of internet services.

The parents in rural areas are not sufficiently educated to be able to properly articulate what exactly are the problems or symptoms their children are facing. They may not also be able to remember all the symptoms due to the number of children to look after. Because of these reasons, the descriptions given by the parents are generally insufficient to help a doctor at arriving at a correct diagnosis.

In order to avoid such situations in the diagnosis of autism procedure, we have proposed a framework which takes only one preliminary symptom and then by applying various machine-learning methods, the targeted symptoms can be selected. These symptoms can then be confirmed from the parents. The symptoms are extracted from the electronic health record (EHR) for autism. The preliminary symptoms of various diseases may be similar. In this chapter we have introduced a disease diagnosis technique, which can narrow down to a targeted disease from the preliminary symptom by applying some machine learning techniques. Given a preliminary symptom, this technique pulls the next possible symptom from the electronic health record (EHR). The EHR which contains the previous autistic children's symptom sets is used a database for the confirmation of the disease, which contains the available autism types and it's symptoms set based on domain experts inputs. There is the possibility for many types of diseases if there is only one symptom but if we make a symptom set by following step by step procedure and then confirm from the parent, then it may be a reliable diagnosis procedure to detect an autistic child. The challenge in this scenario is to find the most correlated symptoms one by one, which is to be confirmed from the parent and thereby proceed to the targeted disease. We have used the Association rule (AR) for the association of the known preliminary symptom with the possible correlated symptoms. Then the maximum relevant symptom is found among all the other co-related symptoms. The Mutual Information Difference (MID) rule is used to select the targeted symptom based on the mutual dependency of the previously selected two correlated symptoms. The same process is applied to find the second maximum relevant symptom and so on. Many times due to social taboos also parents do not want to express the symptoms of their children who might be autistic. We feel our system will help the doctors for arriving at a more accurate diagnosis, and also help the rural parents in articulating the symptoms of their children in a better way.

2 Related Work

The application of IoT in healthcare domain improves the human life expectancy. The application of Machine learning techniques in health domain makes diagnosis easier and faster. The mental and physical status of the human life can be improved because of advanced healthcare [1]. In case of emergency, the sensors send alerts to the care givers. The different types of automation systems improves healthcare by using WBAN. Automation needs different types of sensors and uses machine learning techniques for alert system [2]. The alert system saves lives by monitoring hypertension, fitness and send alert to the care provider in different emergency situation. Autism is a neurological developmental disorder, the child suffering from this disease generally have limited social interaction and communication, known as social phobia [3]. Autistic children can regain their social and communication skills by the help of therapy and regular practices of activities [4]. The early detection of the disease by a sequential rule mining process is explained in [5], and it uses the clinical database for mining the information. It uses the classification modelling for the categorization of the disease type. The "reach-and-throw" techniques for autism identification has developed by using Support Vector Machines (SVM) [6]. The group of classes can be formed by using SVM, and it uses the quadratic programming with the linear constraints [7, 8]. A skin conductance sensor measures the galvanic skin conductance, which is used to identify the autism in a child in the very beginning of birth week, which is explained in [9]. We have gone through many disease diagnosis procedures which use machine learning, but all the procedures use the detailed symptoms of the patients for the identification of the disease. We have taken the scenario of identification of autism in children of rural areas, where the parents are not able to express all the symptom of their child. To overcome this problem and for easy diagnosis of autism in rural children, we have proposed a framework based on the association rule of machine learning [10] and also mutual information dependency rule for the selection of the most targeted symptom.

3 The Proposed Framework

Our proposed diagnosis system needs only one preliminary symptom to arrive at the predicted disease. The proposed framework is given in the Fig. 1. The initial symptom pulls the next possible symptom from the EHR by applying Association rule and then the highest possible symptom will be selected and confirm from the parent. After each confirmation the symptom set will be checked with the Domain Expert Knowledge Database (DEKD). If any matching disease is found with the same symptoms set

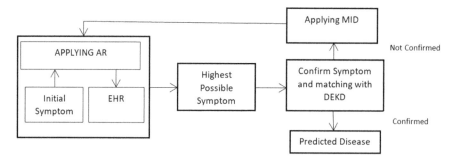

Fig. 1 Overview of our system framework

then the process is stopped, else the mutual information dependency (MID) rule is applied for getting the further targeted symptom. The previously selected symptoms are mutually dependent on the targeted symptom. The process is repeated until getting the confirmed disease from the DEKD, else the process shows "disease not found".

We have proposed an algorithm (Mutually Associated Information, MAI) given below, which assists the doctors to diagnose affected individuals. It also helps parents to recall and express properly the symptoms which are experienced by their children. The Machine learning Association Rule (AR), along with Mutual Information Dependence (MID) rules are applied to extract the targeted symptoms set, which will be confirmed from the parent for the predicted disease. We used Electronic Health Record (EHR), which contains the previous autistic patient's symptom sets for different types of autism. The targeted symptoms for the preliminary symptom will be pulled from the EHR. We used Domain Expert Knowledge Database (DEKD) for the confirmation of the symptom set of the disease.

3.1 Extraction of Symptoms Through AR and MID Rule

Using AR rule, it is easy to find out the other co-related symptoms. Only those symptoms which are relevant to the initial symptom would help a physician to diagnose easily. The association rule will help to extract the relevant symptoms from the EHR. The association rule of machine learning is explained as follows:

According to the Association Rule (AR) of machine learning:

$\rightarrow Y$

$\rightarrow$ Patient initial symptom identified from an unknown set S
$Y \rightarrow$ The target symptom to be pulled from a known symptom set

Y is present in the domain knowledge database, may be because it was earlier experienced by any other patient or was added as domain knowledge.

In order to find Y we rank the rule in the form of 'interestingness' at that particular time period.

'Support' and 'Confidence' are two factors of interestingness.
Confidence (Conf) of rule→Y
Probability of Conf(→Y) =
Probability of Support of is
Support () = Number of time '' experienced

According to the Mutual Information Dependency (MID) rule, the two symptoms are mutually dependent on another symptom which may form a set of symptoms for a predicted disease.

For Example, and form a set of symptom and makes a set of symptom for a disease but according to the Mutual Information Dependency (MID) rule, the two symptoms like and are mutually dependent on. The symptom set will be { }.

4 Mutually Associated Information (MAI) Algorithm

Input: EHR of autistic patients (), disease (, (i=1 to n), X: Preliminary symptom, Y: Targeted Symptom
Output: Identification of confirmed set

Step 1: **Find** all conf (X→) from to // association rule (AR)
Step 2: **Find** Y = , maximum of all conf (X→) //confirm about Y from parent, if not 2nd highest and so on
Step 3: X = {X, Y}
Step 4: **If** X is in , **then**
Step 5: **go to** step 10
Step 6: **Else**
Step 7: **Repeat** step1 to step 4 // Mutual Information Dependence (MID)
Step 8: **Until** X matches with (i=1to n)
Step 9: **End If**
Step 10: Confirmed type of autism.

Suppose the autism patient symptoms sets are abbreviated by the following letters for the diseases:

Let, are the different types of autism are, -- and symptoms are,, --- of the autism(according to domain expert).

Let the *n* types of autism symptom sets are:

= {, , },

={ },

={, }

= {, , }

= {}

Suppose the parent is expressing only one symptom of the child is''.
Considering support of symptom '' is 13.

By applying association rule (AR) we get,

'' experiencing 11 times with

Confidence of → = 11/13= 0.84 i,e in EHR , likewise

'' experiencing 1 time with

Confidence of → = 1/13 = 0.07

'' experiencing 8 times with

Confidence of → = 8/13 = 0.61

'' experiencing 2 times with

Confidence of → = 2/13 = 0.15

'' experiencing 2 times with

Confidence of → = 5/13 = 0.38

'' experiencing 2 times with

Confidence of → =2/13 =0.07

Maximum confidence value i.e. → = 0.84

After confirming '' from parent

We take and as initial symptoms
Applying MID, it is shown in Fig. 2 how mutually select symptom

Fig. 2 Mutually selecting symptom

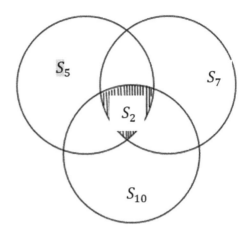

Confidence of () → = 0.6

Conf ( ) → = 0.23

Conf ( ) → = 0.5

Conf () → = 0.1

Conf ( ) → = 0.05

→ = 0.1

→ = 0.4

→ = 0.23

we got maximum confidence →

symptom set matching with domain expert knowledgebase

{} data set is not sufficient for any disease identification.
Maximum confidence occurs in ->, and the highest relevance symptoms are given in Fig. 3.
The data set {} predicted as d7 type autism.

5 Result Analysis and Discussion

Our proposed diagnosis system can be used for any type of disease but this diagnosis system is best suitable for autism diagnosis because, generally the diagnosis procedure of autism does not depend on any blood test or any test report. The diagnosis procedure of autism is based on the child behavior, activities and the parent's opinion. Table 1 contains a part of database of an autistic patient, which we have taken for the confirmation of our diagnosis process. This experiment used EHR, which is collected from the Autism Therapy Counselling and Help (CATCH, Bhubaneswar,

Symptom	possibility of target symptoms	occurrences	selected symptom
S_5	--S_7--	0.84	$S_5 S_7$
	--S_{12}--	0.07	
	--S_{10}--	0.61	
	--S_9--	0.15	
	--S_2--	0.38	
	--S_{15}--	0.07	
$S_5 S_7$	--S_{10}--	0.6	$S_5 S_7 S_{10}$
	--S_2--	0.23	
	--S_{15}--	0.4	
	--S_9--	0.1	
	--S_{12}--	0.05	
$S_5 S_7 S_{10}$	--S_{15}--	0.1	$S_5 S_7 S_{10} S_2$
	--S_2--	0.2	
	--S_9--	0.01	

Fig. 3 Maximum occurrence of symptom selection procedure

India). This diagnosis process has used the DEKD which is having 10 types of autism conditions and having 50 symptoms. We have tested our system with 200 autistic children from CATCH and compared with the real diagnosis system with the same children. Our system only takes a preliminary symptom from the autistic child for diagnosis. During real diagnosis time, the accuracy percentage also varies because of lack of data from the children or parent. We have tested with the autism type Intellectual Disability (ID) with 50 autistic children, in which we found the accuracy rate of 95%, where as 86% was obtained from the actual diagnosis process. Likewise, we tested with the autism type Asperger's Syndrome (AS) with 30 autistic children, in which we found the accuracy rate of 79%, whereas 70% was obtained from the actual diagnosis process. The accuracy rate of 80% in Vagus Nerve Disorder (VND) was found with 30 autistic children, whereas 86% was obtained from actual diagnosis process. The accuracy rate of 88% in Multiple Sclerosis (MS) was found with 30 autistic children, whereas 90% was obtained from actual diagnosis process. The accuracy rate of 70% was found in Pervasive Developmental Disorder (PDD) tested with 30 autistic children, whereas 85% was obtained from actual diagnosis process.

Finally we got the average accuracy of 83% from our proposed system. We have shown one real test case of Intellectual disability in the Fig. 4. Communication disorder with stress is taken as the initial symptom of the child and reached to a predicted disease as Intellectual disability. We have used arrow mark ($\rightarrow$) for positive symptom and × mark for the negative symptom. The next targeted symptom is pulled on the basis of positive symptom which is also confirmed from the parent. The Fig. 5

IoT in Autism Detection in Its Early Stages

Table 1 Database for symptom set by domain expert decision

A	B	C	D	E	F	G	H
Autism Disease Name	Symptom-1	Symptom-2	Symptom-3	Symptom-4	Symptom-5	Symptom-6	Symptom-7
Akathisia	Anxiety	Neuropathic pain	Low iron levels	Rheumatoid arthritis	Stress	Insomnia	
Alice in wonderland syndrome	Migraine	AIWS	Perceptual distortions	Micropsia	Spatial perspective	Memory loss	Lingering touch
Anhedonia	Social withdrawal	Flat affect	Depressed mood	Anxiety	Social anhedonia	Prodrome	
Bruxism	Attrition	Hypersensitive teeth	Glossodynia	Trismus	Headaches	Periodontal ligament	
Capgras delusion	Prosopagnosia	Emotions	Autonomic arousal	Reduplicative paramnesia			
Catalepsy	Rigid body	Rigid limbs	Slow breathing	Parkinson's disease	Epilepsy	Cocaine	
Cataplexy	Muscle paralysis	Hypnagogic hallucinations	Emotional reactions	Embarrassment	Double vision		
Circumstantial speech	Slowed thinking	Invariably talks	Word salad				
Confabulation	Unaware of the accounts	Inappropriateness	Lack of memory	Verbal confabulations	Behavioral confabulations		
Intellectual disability (ID)	Stress	Communication disorder	Oromotor	Aphasia	Dislexia	Discalculia	Vitia battery
Munchausen syndrome	Seizures	Bleeding	Poisoning	Apnea	CNS Depression	Repeated infections	Rashes

(continued)

Table 1 (continued)

A	B	C	D	E	F	G	H
Autism Disease Name	Symptom-1	Symptom-2	Symptom-3	Symptom-4	Symptom-5	Symptom-6	Symptom-7
Palinopsia	Migraines	HPPD	Trazodone	Idiopathic	Head trauma	Mirtazepine	

IoT in Autism Detection in Its Early Stages

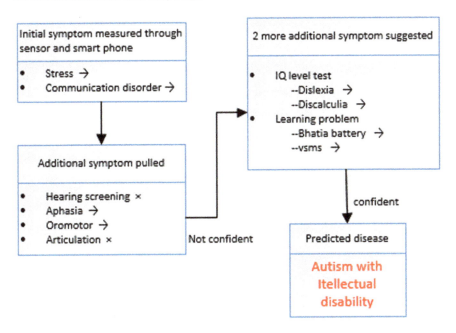

Fig. 4 Intellectual disability tested through proposed system

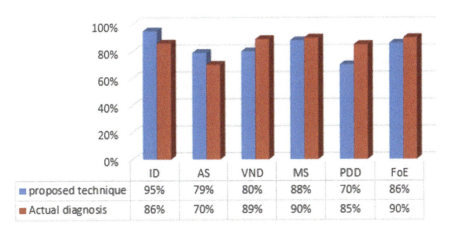

Fig. 5 Comparative graph for proposed method and actual method of diagnosis

shows the comparative graph of the above mentioned 6 types of autism diagnosis with our system and with the actual diagnosis procedure.

6 Conclusion

The AR and MID rule of machine learning are used to extract the targeted symptoms from EHR. The EHR data is accessed from the cloud. The various devices which measure past patient records are also connected to the EHR. They send their data to the EHR for storage and later retrieval. So the internet connectivity should be there among all the interconnected devices, to access the updated EHR for the diagnosis process. We got the average accuracy of 83% by testing different types of autism. This diagnosis system assists a doctor, for predicting children with autism in rural areas, by prompting with sets of associated symptoms.

References

1. J.A. Salomon, H. Wang, M.K. Freeman, T. Vos, A.D. Flaxman, A.D. Lopez, C.J. Murray, Healthy life expectancy for 187 countries, 1990–2010: a systematic analysis for the Global Burden Disease Study 2010. Lancet **380**(9859), 2144–2162 (2012)
2. D. Apiletti, E. Baralis, G. Bruno, T. Cerquitelli, Real-time analysis of physiological data to support medical applications. IEEE Trans. Inf Technol. Biomed. **13**(3), 313–321 (2009)
3. American Psychiatric Association, *Diagnostic and Statistical Manual of Mental Disorders (DSM-5®)*. American Psychiatric Pub (2013)
4. M. Helt, E. Kelley, M. Kinsbourne, J. Pandey, H. Boorstein, M. Herbert, D. Fein, Can children with autism recover? If so, how? Neuropsychol. Rev. **18**(4), 339–366 (2008)
5. Y.T. Cheng, Y.F. Lin, K.H. Chiang, V.S. Tseng, Mining sequential risk patterns from large-scale clinical databases for early assessment of chronic diseases: a case study on chronic obstructive pulmonary disease. IEEE J. Biomed Health Inf. **21**(2), 303–311 (2017)
6. P. Perego, S. Forti, A. Crippa, A. Valli, G. Reni, Reach and throw movement analysis with support vector machines in early diagnosis of autism, in 2009 Annual International Conference of the IEEE Engineering in Medicine and Biology Society (IEEE, 2009), pp. 2555–2558
7. R. Begg, J. Kamruzzaman, A machine learning approach for automated recognition of movement patterns using basic, kinetic and kinematic gait data. J. Biomech. **38**(3), 401–408 (2005)
8. J. Wu, J. Wang, L. Liu, Feature extraction via KPCA for classification of gait patterns. Hum. Mov. Sci. **26**(3), 393–411 (2007)
9. B. Nehme, R. Youness, T.A. Hanna, W. Hleihel, R. Serhan, Developing a skin conductance device for early Autism Spectrum Disorder diagnosis, in 2016 3rd Middle East Conference on Biomedical Engineering (MECBME) (IEEE, 2016), pp. 139–142
10. B.L.W.H.Y. Ma, B. Liu, Y. Hsu, Integrating classification and association rule mining, in Proceedings of the Fourth International Conference on Knowledge Discovery and Data Mining (1998), pp. 24–25

Significance of IoT in Education Domain

Hrudaya Kumar Tripathy, Sushruta Mishra, and Krushnakanta Dash

1 Introduction to IoT

With recent technological advancement, Internet of things (IoT) is emerging as a global computational network where everything and everyone are supposed to be connected to Internet [1]. With constant rise in the number of devices being connected to Internet, knowledge acquisition becomes simpler and reliable. Interactions among intelligent objects are a breakthrough technology, but the technology enabling IoT is not a new thing for mankind [2]. In simple words, IoT can be described as a concept where data aggregated from various types of sources or things can be gathered to a virtual environment on an existing Internet architecture [3]. The main idea of IoT is allowing useful information exchange in an automated manner between distinct real-world entities in our surrounding backed by advanced technologies like wireless networks and radio-frequency identification.

Different kinds of devices are connected by IoT which include laptops, smartphones, tablets and hand-held systems. Some others are equipped with devices that measure blood pressure, heart rate, devices for pet animals, autonomous vehicles or home appliances, etc., with the help of different sensors, these devices collect relevant information and send to other processing devices for efficient decision making. Majority real-life applications that we come across at present are smart in nature but

H. K. Tripathy · S. Mishra (✉)
School of Computer Engineering, KIIT (Deemed to Be University), Bhubaneswar, India
e-mail: sushruta.mishrafcs@kiit.ac.in

H. K. Tripathy
e-mail: hktripathyfcs@kiit.ac.in

K. Dash
IBM, Bhubaneswar, India
e-mail: krusdash@in.ibm.com

© The Author(s), under exclusive license to Springer Nature Singapore Pte Ltd. 2021
S. Kumar Pani and M. Pandey (eds.), *Internet of Things: Enabling Technologies, Security and Social Implications*, Services and Business Process Reengineering,
https://doi.org/10.1007/978-981-15-8621-7_6

they fail to exchange information with each other. Thus, enabling these devices to share information with each other will develop a variety of innovative applications [4]. IoT with its innovative applications enhance the quality of living in society. Few challenging issues associated with IoT are privacy and security, mobility, availability, scalability, management and trust [5]. Based on different needs of users, there exist various applications of IoT like smart industry, smart retail, smart agriculture and transportation [6]. Figure 1 illustrates common applications of IoT in modern times. The influence of IoT helps organization in decision-making process regarding learning process, operational effectiveness and inside campus security. Its ability to keep track of students, objects and other staffs, thereby connecting devices across boundaries determines a higher safety level to institutions.

Among various smart applications supported by IoT, smart education driven by digitalization and virtual learning is a chief element which can revolutionize society. In a smart city environment, smart education is a very important factor. Figure 2 shows

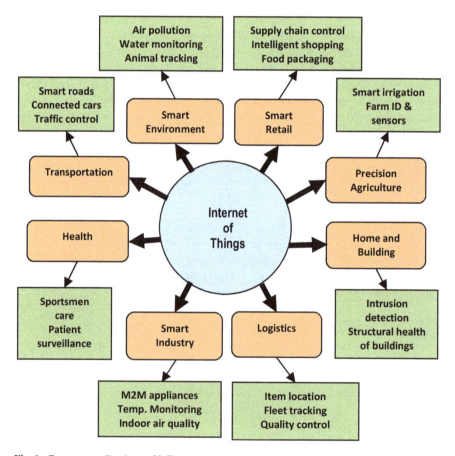

Fig. 1 Common applications of IoT

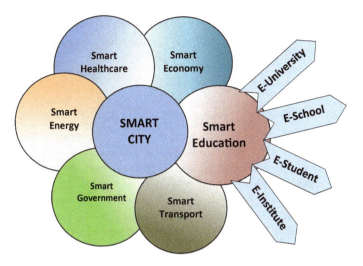

Fig. 2 IoT smart city segmentation

a very simple yet effective classification of smart city segmentation. As seen in the figure, smart education forms a very crucial element of smart city division which is implemented by IoT. Through IoT, the entire education domain has witnessed the transformation. The impacts of IoT are visible in educational institutions, schools, universities and day-to-day lives of student. Few smart implication of IoT in education includes E-institution, E-school, E-education and E-student.

This chapter is sequentially organized in a very systematic manner. The first section introduces the IoT technology to the audience and readers. The concept of IoT, its application zones and its relevance in modern society is highlighted in this section. The second section stresses upon the role of IoT technology in revolutionizing education field. It signifies the role of IoT concept in various aspects and functionalities of educational institute's campuses. Various vital issues and challenges faced by IoT technology in educational domain are represented in the third section. The fourth section lists some important application examples of educational objects that make use of IoT concept in modern times. Various substantial benefits of IoT in education are presented in the fifth section of this chapter. The sixth section discusses various real-time application frameworks that utilize IoT technology in educational premises. Some of these applications include waste management, learning methods, sports coordination, students and staffs security and many such areas. An application of IoT concept in E-learning process is being highlighted in seventh section. An opinion survey of the impact of IoT concept in modern day education is presented in the eighth section of this chapter. Views of several computer engineering youths were taken into consideration for this analysis. It is observed that IoT has a positive influence among youths and majority of them preferred IoT technology-based education over traditional teaching–learning process. The ninth section presents a real-life application of a flipped classroom concept using IoT which was implemented in a private university in Odisha. The final section concludes our chapter, thereby inferring that IoT-based education provides a clear edge over conventional way of education.

2 IoT in Education

Significance of technology in the education sector is very vital in developing a connection, and thereby educating students at different level. From the use of innovative and creative teaching–learning process to the use of tablets in classroom, education has revolutionized the lives of people. However, these transformations are less when compared to the mass diversities that the educational domain is witnessing due to the emergence of IoT technology. IoT has a positive and endearing impact in the field of education. Its impact is visible not only in teaching–learning process but also in the educational institutional infrastructure [7]. Educational IoT is represented as two faceted due to two reasons. Firstly, IoT can be implemented as a technological agent in enhancing academic infrastructure. Secondly, it can be introduced as a core subject domain to teach and learn core fundamentals of computer science [8]. IoT technology is improving the educational sector in all aspects which include school, college and university level of teaching. IoT can be beneficial to all aspects of education from classroom to entire campus and from student to faculties. Nowadays, in reputed educational institutions, IoT is used as a research-oriented course and helpful in developing innovative real-life application projects. IoT attracts students of all levels due to its exciting and stimulating aspect. It is an ideal platform to learn concepts of computer science [9]. With the availability of mobile technologies, IoT can help educational societies to track various resources associated with education and students. IoT plays an active role in not only teaching and learning but also in overall assessment process of its implementation.

The understanding and interpretation of IoT are fruitful in resource delivery in a creative way dealing with the audiences. Thus, IoT is capable of impacting detailed aspect of education process of students. This perspective of IoT allows various stakeholders with a dynamic view of staffs, participants and resources. It thereby is helpful in effective decision making, automated way of executing and having various security and privacy characteristics. Few developments associated with IoT in the forthcoming days are presented in Table 1. IoT in education is generalized with four pillars that form the basis structure. These generalized pillars of educational IoT include process, people, data and things and this is represented in Fig. 3.

IoT creates a new picture of education when it is integrated with modern technologies like data analytics and user mobility. IoT enables educational institutions to:

- Develop innovative ways for students to learn by providing more adaptive and personalized learning methods like game-based learning.

Table 1 Paradigm shift to IoT in future education

Current state	Potential with IoE
Physical attendance in presence of faculties	Scale teachers and best quality of instruction-any device anywhere
One-time instruction in one location	Scale content recordable and replicable instruction, any time, any venue
Static, linear content with low control	Learn at your own pace, focus on relevant content only, richer, interactive content
Costly instructional resources	Access to crowd-sourced content, ability to customize curriculum
Ad hoc decision making	Data-driven decision

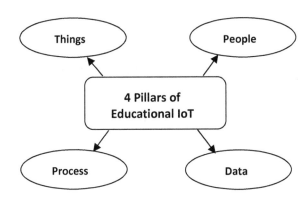

Fig. 3 Pillars of IoT in education field

- Change the manner in which faculties deliver their chapters and validate its effectiveness by the help of smart audio-visual devices and online testing.
- Simplify functionalities of management of institutions by monitoring vital infrastructure and developing efficient and less costly methods for HVAC and landscape management.
- Create a safe environment for faculties and students with the provision of digital surveillance cameras with smart door locks inside campus.

Several works have been undertaken with respect to role of IoT in educational field. A real-time IoT-based system model was developed and implemented in University of Padova [10]. A Web-based service model using wireless sensor network was developed for validation of faculties'–students' programs. In [11], a detailed study was discussed where the effect of distinct technologies like cloud platform, IoT and machine learning are studied in relation to distance education. A research work was carried out in [12] to describe the significance and role of IoT associated with cloud computing in the field of education which also distinguishes smart campus from traditional digital campus. In [13], an integrated IoT-based architectural framework was proposed and presented in an academic set up by a group of faculties. Chin and Callaghan [14] stress the importance of IoT and its role in imparting fundamental core concepts of many programming languages to students. In another research work [15], IoT was applied to facilitate in development of an interactive system framework to teach English language. It used visual and voice sensors to rectify the English pronunciation of participants. A research work [10] utilized objects associated with identity tags and knowledge coordination model to gather information while process students' learning technique with the use of learning-based analytics methodologies.

3 Challenges of IoT Deployment

The IoT technology presents massive flow of information posing management and functional challenges to the network framework along with enhanced privacy issues from all end points. These issues need to be addressed properly by network administrators at respective educational sites. Cost-effective network architecture is required at schools and universities level to deal with this data flow. Some critical criteria that the infrastructure must meet are discussed here.

Provide a Simple, Automated Process for IoT Device on Boarding

Huge IoT systems comprises of several thousands of sensors. Hence, monitoring these endpoints is complicated and chances of error are high. Automating the system makes the network more dynamic and devices possess self-recognizing capability and are allotted to suitable network.

Supply the Correct Network Resources for the IoT System to Run Properly and Efficiently

Several devices in IoT model provide task-critical information that demands a high level of quality, for example, a few educational task need predefined bandwidth requirements on a high performance network framework to maintain high reliability and data delivery.

Provide a Secure Environment Against Cyber Attack and Data Loss

IoT tasks require huge amount of data transfer in real-time environment. Several sensors are embedded in the system with the presence of many end points. Hence, security is a major issue that must be taken care of. At every level of data flow, security is a requisite. A minor breach in security might disclose personal information of any student related to his personal data, family financial status or his medical record.

Reliable Wi-Fi Connection

Continuous Internet connectivity is very critical for IoT-based applications, especially in education domain. Very high speed Internet connection is required for efficient real-time audio and video streaming of data.

Management

There are few system models which are incompatible and may create an obstacle to develop an IoT infrastructure that is both available and reliable to all users. For IoT to be successfully implemented, an organization should take care that its teaching process as well as its IT devices support the usage of IoT in classroom campus.

Cost

The entire infrastructure set up of an educational framework can be very costly. Hence, to limit the cost within a specified range is a major limitation in setting IoT supported set up.

4 Real-Time Smart Education

With the use of IoT technology, universities and schools are getting smarter and more progressive. The potential of IoT lies in redefining the manner in which faculties, students and other staffs interact, share and interconnect to devices to meet technology demands, thereby helping in enhancing learning process and reducing educational costs. Some vital and popular examples of application of IoT in educational domain include the following.

- Smart whiteboards and interactive digital devices to collect and aggregate information for students and faculties to be used in classroom or outside, thereby enhancing instruction control and increasing learning outcomes.
- Smart classrooms well equipped with IoT controlled things to enable for remotely monitored. It enables to set the classroom well ahead of time of lecture delivery. The faculty can reset the lighting, overhead projector and reset the room temperature accordingly, thereby saving time and power. The device to be monitored should be capable in receiving orders to initiate a specified task while they are to be embedded with sensors.
- IoT-based solutions like smart temperature sensors and smart heating, ventilation and air conditioner device to decrease power consumption and automate operation management.
- Smart ID cards of students, attendance monitoring devices, college bus tracking systems and vehicle parking sensors to monitor the physical location of students.
- Connected surveillance cameras, wireless door locks and facial recognition systems to provide surety and privacy for faculties, students and staffs.
- Research workshops and programs along with modern automated models in vital areas like agriculture, industry and health care.
- Intelligent advanced technologies like big data, learning analytics, cloud computing, Internet of things, etc., promote materialization of smart education.
- Smart parking regions well embedded with sensors to monitor the availability of parking areas, thereby providing indications of available area capacity to save power and time.

- Smart secure environment embedded with camera sensors to provide alert messages to a secure monitoring room to enhance security inside building area and prevent irrelevant events.

Smart feedback of student to permit students in participating in learning acquiring system enhancement by the help of a sensing mobile application.

5 IoT in Education Advantages

Better Learning Experience

Since digital learning consists of a number of smart devices, majority of time is wasted in device monitoring and decision making. In IoT platform, management of interdevice communication is performed and so maximum learning experience is provided to students.

Improved Operational Efficiency

There are many stakeholders involved in educational institutions. Students, resources and associated staffs must be tracked properly to get the desired result. Management of devices and its operations are to be done intelligently by implementing IoT technology. Efficient operational management guarantees the overall success of the system. It has many constituents involved. The overall control of individual component is a successful method to uplift the system's effectiveness.

Reduced Cost

Proper management of several control units in an institute reflects the major expenditure in the organization. The automated communication occurring among all the institutional units denotes the overall costs incurred. As detailed observation will be carried out in an automated fashion, it leads to the decrease in the expenditure.

Reliability Concerns

A reliable application leads to higher outcome. The individual parts which are available in the entire system represent its reliability. Since IoT deals with management of individual components, it denotes the reliability.

Safety Considerations

IoT supported system model is fruitful in tracking the safety and security issues of an educational organization. Various security concerns related to entry of external persons, movement of students, fire safety and similarly other aspects can be monitored effectively. Since communication between several devices becomes automated, so specified surveillance system can be installed inside campus. It is helpful in managing outdoor security campus by the use of wireless objects. Similarly, the management of various vehicles along with their safety can be efficiently provided.

6 IoT-Based Smart Campus

The main idea behind smart environment is that smart surroundings and smart equipments are everywhere available and it is there for all to perform their day-to-day tasks. Smart offices, smart homes and other smart regions all form the part of smart environment. The basic objective of IoT-oriented smart environment is to enable simplicity in daily routine works. Let us take an example of driving cars where we must be aware of optimal route and traffic conditions. This information can be obtained by the use of sensors and smart devices in cars. Reasoning-based learning and prediction are the main objectives of smart environment.

Thus, a smart environment can be represented as an environment which can acquire and use knowledge about the environment, thereby enhancing their overall experience. Generally, school and college campus gets very reliable Internet connection. Every campus comprises several entities such as doors, windows, projectors, classrooms, printers, parking area, laboratories and building. If specific sensors and QR tags or any other related IoT technology are used to these entities, they will be transformed into smart objects. Hence, a smart IoT-oriented campus acts as an amalgam of several distinct smart objects embedded within one single framework. An IoT-based smart campus includes objects like:

- IoT-based E-learning facility
- IoT-based classroom and laboratory
- IoT-based sensors for sharing of notes
- IoT-based sensors for mobiles applications
- IoT supported in-campus hotspot.

Figure 4 illustrates some common implementation of smart campus in educational organizations. These include vital constituents such as smart vehicle management, smart sports, smart classrooms, smart laboratories and smart health by wearables.

6.1 Smart Vehicle Management [16]

Effective tracking and management of school and college buses can be done with the use of smart technology. Following includes some vital aspects of smart vehicle control.

- *Bus Attendance*: Attendance of students is regularly updated on cloud service with the usage of data collected from RFID reader.
- *SMS Alerts*: Parents of students are sent picking and arrival alert messages regularly in automated fashion.
- *Route Adherence*: The bus driver receives alert messages when it goes in wrong route or the fuel capacity is low, thereby ensuring safety.

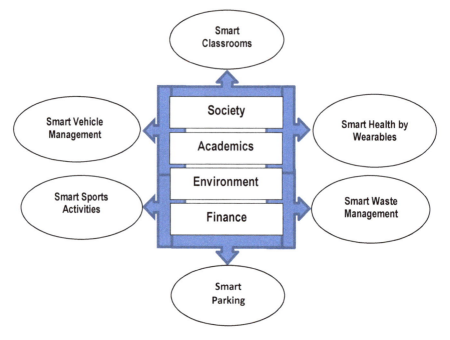

Fig. 4 Some constituents of smart campus

- *Emergency Management*: In scenario of any emergency case, the bus travelling route is selected and student information like blood group and contact details are immediately sent to the nearby hospital for safety purpose.
- *Live Tracking*: The complete travelling of bus can be viewed on map providing more safety to the persons inside bus.
- *Voice Call*: The bus driver is well equipped with emerging voice call facility, thereby receiving information from institution authorities via hands-free communication.
- *Driver Behaviour*: Automated reports are provided regarding driver's behaviour. Accordingly, the driver will be rated based on driving habits, timeliness and other factors.
- *Preventative Maintenance*: Applications and sensors avoid mechanical related failures by sensing alert messages when a problem arises.
- *Download Documents and Dashboards*: Documents like location of bus, attendance details and speed with other related trends can be accessible anytime and anywhere.

6.2 Smart Parking

Traffic congestion due to heavy traffic is a major concern nowadays in many educational places. Hence, smart parking mechanism should be facilitated in such places for effective vehicle management of staffs and other people. Staffs and students waste time in finding a suitable parking slot in a busy university area and this leads too rise in carbon emission level. Thus, this parking problem must be addressed for the overall management of the students and university, thereby reducing problems for university. This ultimately decreases the effectiveness of all parties involved which result in heavy traffic in the campus. It generally occurs during peak busy hours and may get worse if proper steps are not taken. Hence, there is a need of smart and reliable yet simple parking coordination mechanism in these campus. Here, an easy smart parking model is discussed which can be implemented in educational institutes. The model provides an IoT concept implementation in deploying a sensor network in a smart campus building.

The discussed model utilizes the sensor nodes arranged at suitable vehicle parking places to identify and sense the availability of parked vehicles. Subsequently, the information will be sent to a microcontroller which is responsible for transmitting those sensed data to an online repository using wireless medium. The repository acts as an interface between mobile application and sensor nodes, which thereby gather all sensor data.

This mobile application uses the information available in server to identify and locate the free parking zones. Hence, this will enable the users to access and manage real-time data from any desired place. A sample smart parking model is shown in Fig. 5.

6.3 Smart Classroom

The concept of smart classrooms refers to an automated and intelligent environment well equipped with modern learning needs related to recent technology or smart things. These smart objects include cameras, sensors and microphones to determine satisfaction level of students related to learning task. Smart devices are crucial in providing ease and comforting environment in managing classes. It thereby provides a much better teaching and learning environment.

Smart Classroom Management

An approach used by faculties to control a classroom is referred to as classroom management. Due to the presence of smart devices, faculties can decide when to increase his voice to retain the interest level of students. The application of IoT technology is a latest emerging trend for learning and teaching process that provide a creative approach in handling classrooms and overall educational process. Some general IoT devices used inside classroom are:

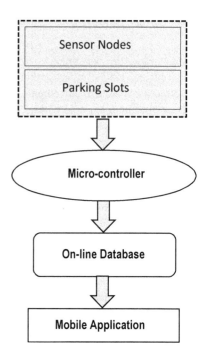

Fig. 5 Flow chart for smart parking

- Interactive smart whiteboards
- Mobile objects and tablets
- E-books
- ID cards of students
- Temperature and humidity sensors
- Surveillance cameras
- Electric lighting devices and maintenance tools
- Automated attendance tracking model
- Wireless door locking system.

Faculties can determine the learning level and requirements of students through a smart classroom and this can benefit both students and faculties. Most importantly, IoT-based classrooms enable participants to interpret actual objective of the use of this concept to make the learning task relatively easier. This advancement in technology in educational domain enables faculties to develop classrooms which are more useful, collaborative and productive through IoT technology. A pictorial representation of smart classroom is depicted in Fig. 6.

Smart Classroom Attendance System

Task of attendance call can become very hectic and consumes time. An IoT-based automated attendance system can be helpful to save time and effort. Various IoT-based smart attendance systems are developed to gather and monitor attendance of students at regular intervals in a precise manner. An example includes use of RFID tags which

Fig. 6 Smart classroom [17]

are attached to ID cards of students. The smart classroom roll caller system can be configured in classrooms which can be used to record identity of students regularly. A LED display may be used to show the attendance of all students after the attendance task. The copy of attendance of every student is maintained in management office. In some institutes, a Web-based attendance management model is incorporated with the use of NFC technique applying android smartphones. The attendance of students is tapped and is automatically saved in the server. Both faculties and students can use their smartphones to verify the presence of students in their class.

Real-Time Feedback on Lecture Quality

The quality of lecture is directly related to understanding level of students. The feedback of students plays an active role in enhancing quality of lecture and overall teaching process. An IoT-based system model was developed that can sense and monitor reactions of students to a particular lecture. A real-time feedback system can be developed by the use of IoT technology which can help both students as well as faculties in coordinating class flow.

6.4 Smart Laboratory

A smart comprises innovative hardware, online lesson libraries and regular professional development modules. Some of the basic purposes of smart laboratories are as follows.

- To enhance learning and academic skill level of students.
- To motivate students and staffs.
- To match the requirements of faculties and students.
- To improve results for both institution as well as individuals.

A smart laboratory is well equipped with several advanced types of equipment that are interconnected with each other. They are able to transmit and receive information simultaneously at any time with the help of sensors and actuators. Students and faculties' laboratory monitoring kit with a series of sensors with Arduino boards are available to facilitate wireless means of communication in the laboratory and can engage themselves to find their specific information at any time. A sample smart laboratory is shown in Fig. 7.

Fig. 7 Smart laboratory (IoT) demonstration [18, 19]

6.5 Smart Sports Activities

Now, sports are not just about the game. As our world is getting developed into the digital world so it is forced to the sports industry to adopt it, and the Internet of things (IoT) helps to connect the gap between the physical world and the digital space. Figure 8 denotes certain criteria for smart sports. To address a particular need or challenge, today, many educational organizations are using IoT in specific areas. On three main areas of sports, where teams may focus their activities with the support of different IoT efforts:

Player development: Training for coaches facilitates, player management, and address key situations in each game is revolutionizing by IoT. Vast amounts of data can easily be processed by coaches to obtain metrics on player efficiency, and the performance of the player and opponent weaknesses to better develop an in-game strategy by using the combination of advanced analytics with the sensors and game video.

Player safety: IoT is forming the way that sports doctors, physical advisors and group specialists are diminishing wounds and helping players recuperate quicker. Installed gadgets, for example, keen insoles and implicit chips offer constant following that gives an all-encompassing perspective on the competitor, enabling associations to settle on the best choice for their life span and well-being.

Fan engagement: Fan commitment: In smart stadiums, for improving the computerized commitment and eventually the in-field experience at last IoT is utilized. The concept of the future stadium is here, where fans can engage with their favourite teams and athletes which were impossible before. For new stadiums

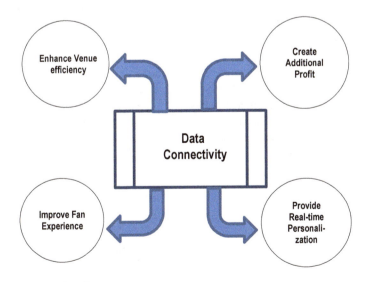

Fig. 8 Smart sports benefits

and stadium improvements, many organizations had already invested billions of dollars which will improve the live experience of fans.

6.6 Smart Waste Management

Irrespective of developed or developing countries one of the main problems nowadays is waste management. The fundamental issue of waste administration is that the trash container in the general population places gets flood before the following cleaning process. Due to these various kinds of diseases are spreading it also leads to bad smell and affects the beauty of the place. The fundamental purpose to study is to build up a savvy trash ready framework that will properly manage garbage management. This study discusses a smart alert system for cleaning the garbage which will give an alert signal to the Web server of the municipal instantly to clean the dustbin with appropriate confirmation based on the garbage level filling. This process is aided by the ultrasonic sensor which is interfaced with Arduino UNO to check the level of garbage filled in the dustbin and sends the alert to the municipal Web server once if garbage is filled. The aid of an RFID tag is used by the driver after the dustbin clean. RFID is a computer innovation that is utilized for the confirmation procedure, and furthermore, it additionally upgrades the shrewd trash ready framework by giving programmed recognizable proof of trash filled in the dustbin and sends the status of tidy up to the server certifying that the work is finished. By an embedded module integrated with the help of RFID and IoT that fertilize, and the entire process is upheld. Municipality authority monitors the status of waste collection in real life with the aid of this system. Moreover, the essential measures could be adjusted by therapeutic/interchange. To intimate the alerts from the microcontroller, an application is developed for android which is linked to the Web server and the urban office. The cleaning process is performed using automated remote monitoring so that it will decrease monitoring and verification manually. The notifications are sent using the Wi-Fi module and the notifications are sent to the application present on the android.

Figure 9 shows the working flow which gives the idea of this framework. Trash has a sensor level that has ultrasonic sensors in the bins and when the threshold level is crossed by the bin, and Via GSM the message is sent to the concerned power with the goal that the concerned authority can clean the dustbin as early as they can. Until the dustbin is cleaned the process gets repeated itself.

6.7 Smart Wearables

In every classroom including student and teacher's hand, because in our time it is the pen and paper, and we experience much of our world through the lens. In education, the system wearables system plays an important role in taking the community to the next level. This technology improves student engagement in education and it also

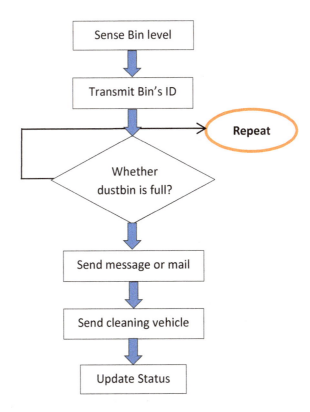

Fig. 9 Flow chart of smart waste management

increases the communication between teacher–student. In the educational domain, several wearable technologies are developed and also implemented. Some of the popular wearables are highlighted in this section.

VR headsets and smart glasses can give understudies top to bottom tactile encounters inside (and outside) the study hall. VR is experimented by many institutes to create "hands-on" experiences to the students in such as medicine, engineering, physics, geography, art and history.

Augmented reality (AR) smart glasses soon smart glasses will also be used by the university classroom. Physical world can be mixed with the virtual world using AR, empowering a gadget wearing the client to take a gander at a genuine article and see pictures and content overlaid on it. In engineering, it is being utilized to gamify complex ideas, making them increasingly available to understudies.

Fitness wristbands are meant to track students' physical activity, development around grounds, crucial signs and resting propensities.

Smartwatches hold a lot of potential for learning improvement. It helps to link directly to a student information system which is present in the institution, wearable can also receive instant notifications from the institution, notices and statements, or it can also be programmed to take or drop call, where students wearing them are naturally enlisted as "present" when they stroll into their study hall or auditorium.

Smart shirts and health sensors for the medical field is extremely valuable tools and it helps to assemble and analysis real-time patient data. Smart shirts use to record biometric data, for instance, breathing, beats, advancement types and UV levels.

Google Glass is used to make the learning procedure smoother and engaging for the student and teacher.

Students can use the glass to—

- Take notes and bookmark significant entries
- View extra-study hall content gave by the addressing educator
- Get a virtual simulation experience of occasions on the athletic field, or in space.

7 IoT in E-Learning

The concept of IoT technology has brought a lot of flexibility in education sector where it is feasible for the teaching and learning process to take place at any time and at any location if faculties and learners are willing to participate. Thus, it brings a great deal of accessibility in educational domain, especially in E-learning concept. By the use of IoT technology in E-learning process, the students can actively participate and interact with the faculties, perform all the assigned tasks remotely, submit the assignments online and getting results real time. Thus, it reduces the dependency on manual procedures. Instead focus can be shifted on the learning process rather than other secondary functionalities. Advanced technologies like RFID and cloud computing can be applied to gather information regarding the learning efficiency. Similarly, faculties can utilize the technology to improve performance of overall education.

Advantages of IoT technology in E-learning system are discussed here.

- IoT helps in connecting the internal education model with the online faculties and learners, thereby accessing massive resources at any time. The faculties can use their experiments and demonstrations to connect to the latest technology in world. Solutions to several queries and questions can be found by linking this huge information content which is stored anywhere and anytime.
- Almost all systems can be interconnected to each other and to the physical world by developing a recent IP address scheme. It helps in generating huge amount of data and interactions among entities. Faculties and students can interact with each other at anytime and anywhere. Human beings in collaboration with robots and software will sort out problems of students from any location. All types of queries can be solved and answered by performing assessment online and yielding related results. This procedure has a positive influence over student's performance.
- IoT eliminates the visible electronic barrier, time restrictions and other obstacles between educators and huge quantity of resources like experienced faculties,

research results and solutions with modern laboratory equipments. IoT with its associated technologies interface this connection. As the obstacles are reduced, eventually, the speed of access to the relevant data is also improved. Faculties and students can access the required data without any delay.
- IoT smart environment has the provision of personalized E-learning interface well equipped with all facilities that satisfies student's needs. Apart from this, IoT smart technology helps to integrate E-citizens into the educational community. It can enhance the E-learner involvement in the learning process.
- IoT is gradually changing the working performance of many E-learning tools and mediums such as wireless network connections, RFID verification and resource dimensions like cloud systems.
- Smart education with the use of IoT technology can be an effective and collaborative E-training platform to several E-communities to collaborate and co-operate competent learning experience, and recent models of E-business and learning process which are required to sustain and maintain the educational decorum. IoT E-learning interface can deal with development of smart cities by uplifting their capability through knowledge and transfer of education ideas.

8 An Opinion Survey of IoT Technology on University Students

Since the maximum consumer of educational IoT is students, it is crucial to interpret the student's needs and requirements along with their expectations form IoT technology. Simultaneously, it is absolutely crucial to determine student's knowledge degree as far as the introduction of IoT concept in education stream is concerned. Hence, a survey was being carried out among the computer science group and a succinct feedback on IoT was summarized. The analysis determines the capability of students to handle the recent technology and their present level of knowledge regarding IoT concepts.

8.1 Experiment Evaluation: Determining Knowledge Levels

Experimental set up

The below hypotheses are grouped and discussed for feasible validation with the use of experimental evaluation.
- H1—Students have a reasonable idea regarding IoT and its applications.
- H2—IoT from student's perspective adds to their learning ability in a positive way.
- H3—Students adapting to digital measures and becoming more IoT friendly in a learning environment.

Table 2 Sample experiment questions in the Likert scale

Questions	Likert question (1—strongly agree, 2—agree, 3—not sure, 4—partially disagree, 5—disagree)
22	IoT facilitates me to make use of recent technology
23	I prefer studying in a campus which supports IoT technology
24	IoT infrastructure can enhance my learning ability
25	IoT-related applications will be more fun to use
26	I would not use IoT due to security issues
27	I prefer my phone to be connected to new learning services on campus
28	I prefer to have an automated lecture attendance recording system
29	I believe IoT can be of help in learning more technologies
30	IoT can enhance the learning capability for me and my friends
31	I believe we are not completely ready to use IoT at present
32	IoT may have a negative influence and create distractions

- H4—A specific use case that understand IoT concept is important.

Hence, the experimental analysis incorporated the following objectives:

- O1—Evaluate hypotheses.
- O2—Determine student's knowledge level concerning IoT.
- O3—Extract student's attitude regarding IoT in educational domain.
- O4—Learn from students regarding the negative impact of IoT technology.

A large scale of students from the computer science department of a university took part in the experimental demonstration. These students are from various undergraduate levels. Also, previously, there was no IoT course being provided to students of the university. A set of queries are highlighted in Table 2. These queries are presented on the Likert scale as depicted in Table 2. Based on the computations, the agreement is stronger when the number is less.

8.2 Experiment Results—Performance Evaluation of Hypotheses

Around 500 students were required to fill up the questionnaire. Table 2 shows the result of few questions relevant to the stated hypotheses and objectives. Mean, mode and standard deviation values are provided here which are relevant to every query presented in Table 3.

Significance of IoT in Education Domain

Table 3 Set of queries with responses on questions

Questions	Mean	SD	Mode	No. of participants for each response (1—strongly agree, 2—agree, 3—neutral, 4—disagree, 5—strongly disagree)				
				(1)	(2)	(3)	(4)	(5)
22	1.52	0.71	1	15	7	3	0	0
23	1.88	1.01	1	11	8	5	0	1
24	1.84	0.85	1	11	7	7	0	0
25	1.8	0.65	2	8	14	3	0	0
26	2.68	0.99	2	3	8	8	6	0
27	1.72	0.74	1	11	10	4	0	0
28	1.48	0.65	1	15	8	2	0	0
29	1.72	0.61	2	9	14	2	0	0
30	1.8	0.71	2	9	12	4	0	0
31	3.04	1.24	4	4	4	6	9	2
32	2.92	1.15	4	4	4	8	8	1

9 Analysis and Discussion

Students have a reasonable idea regarding IoT and its applications.

There are total 14 queries (Q2–Q15) which covers the IoT learning and 5 survey queries (Q16–Q20) analysing the applications of IoT awareness. For the first section, it generated (mean, SD, mode) as (2.25, 1.2) and the second section generated (1.55, 0.67, 1). The results depict the knowledge level of IoT based on application orientation related to accomplished knowledge of selected team of computer science students. This may serve as an input to H4. It is seen that 62% students either agree strongly or agree partially to the IoT concept awareness and 90% students agree for the application section awareness.

IoT from student's perspective adds to their learning ability in a positive way.

Input queries for this analysis involve Q22, Q23, Q24, Q25, Q27, Q28, Q29, Q30 presented queries in Table 2. Output of these queries is shown in Table 3. Approximation of the inferences of evaluation of the rule is (1.72, 0.75, 1). Overall, it is observed that 85% of students either agree or partly agree to the hypothesis. The output also matches the Q30 result as seen in Table 2.

A well-defined use case that interpret IoT concept is important.

Given H1 and the significance of the application area be precise to students, Q28 gives a very unambiguous situation and is beneficial for students. The output of evaluating this query is (1.48, 0.65, 1) that depicts an enthusiasm to adopt such use case.

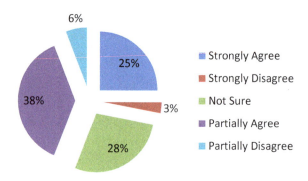

Fig. 10 IoT awareness indicator among experiment participants

As observed in Fig. 10, 92% students either just agree or strongly agree to the query. It denotes digital interest to adopt methods to automate various functions which might be challenging for subsequent generations. In general, the analysis represents that the learning degree of participants is determined to be 62%. It shows that all students are still not aware of the working concepts of IoT technology. The possible reason may be due to lack of teaching of such latest courses in the present syllabus. But it is seen that 90% students have a strong belief regarding the usefulness of IoT to enhance applications in the field of transportation, industry, health care and service. 80% students are of the opinion that proper understanding of IoT technology will enable them to be equipped with better opportunities in the market.

85% of students are of the opinion that IoT technology can increase learning mechanism of them as well as their colleagues. It can thereby be helpful in learning other related technologies too. Students interlink this learning gathering process to the capability of accessing other education-based services with the usage of mobile phones. Hence, they are comfortable to enrol in a campus infrastructure that supports IoT technology. The awareness degree of students about IoT concept is presented in Fig. 6. Still around 14% students are negative regarding IoT with respect to security issues and technology awareness. Applications such as location-based models may create some discomfort, and hence privacy of students is very vital and a reliable technology should take care of such issues. Apart from this, few believe that IoT can create distractions in learning gathering process. Overall survey indicates that more participation at student level towards IoT technology is the scenario at present.

10 Flipped Classroom as Element of IoT Education

An interesting component of educational IoT is flipped classroom which is rarely implemented in developing countries. This flipped classroom is an IoT-based system model where students and participants can gain knowledge by having lectures at their homes and they carry out different assignments related to lectures in classroom in presence of their concerned faculties and educators. All evaluation exams and seminars based on those lectures are conducted in classrooms. Knowledge-based video lectures are developed by educators and are shared among students. Students need to watch these videos before coming to classes for interactive session and evaluation. The faculties can get a precise insight of knowledge gained by students from these video lectures. Application of flipped classrooms helps students in learning any subject at any time and any place at their convenience.

The flipped classroom concept was implemented in Computer Networking course of B.Tech. batch at KIIT University, Bhubaneswar. Three whole sections of third year B.Tech. students participated in the implementation of flipped classroom model. For the first time in KIIT University, this IoT-based model emerged during odd semester of the university for 2016–2017 batch of students. Around 200 students were involved in the flipped classroom concept. Other sections continued their usual education in traditional educational approach. Various experts were involved in preparing video lessons using advanced tools. Based on the knowledge gained, students made small presentations on the topic allotted by educators. Time limit for each presentation was set to 15 min. Educators performed a minor filtration session with the use of IoT service Kahoot and at the end of test students were involved in laboratory assignments with NS2 simulator. During the evaluation procedure, simulator automatically computes marks of students. At the end of semester, evaluation of flipped classroom model was being carried out. The result of survey is depicted in Fig. 11. It was observed that 72% students strongly agree that flipped classroom model was better than traditional lectures. Only 15% were of the opinion that traditional concept was more helpful than flipped concept. 13% students were neutral in opinion.

The implementation of flipped classroom system was analysed by the top management of the university. The results illustrate that students who studied with flipped classroom concept performed much better compared to normal conventional educational approach. Table 4 presents the analysis results. Students who had flipped classroom training comprised almost half of the total students in B.Tech. third year of KIIT University. As it is seen from the table, in every aspect of parameters, flipped classroom-based IoT model performs much better than traditional approach of education. An increase in attendance is observed with flipped IoT model. Similarly, problem-solving ability and self-learning skills are also enhanced with this

Fig. 11 Flipped classroom performance analysis

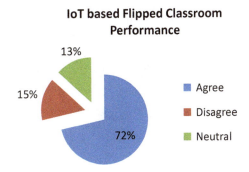

Table 4 Flipped classroom versus traditional classroom approach

Classroom activities	Traditional approach (%)	Flipped approach (%)
Laboratory assignments	67	88
Attendance status	73	93
Quiz evaluation I	76	91
Quiz evaluation II	73	90
Mid-semester evaluation	83	92
Problem solving ability	84	91
Self-learning ability	67	88
End-semester evaluation	76	89

IoT-based approach. Overall, the mean value of all factors is computed and it is seen that with traditional approach and it is only 74.87%, whereas it is 90.25% when implemented with flipping classroom approach.

11 Conclusion

In this era of IoT, it is strongly believed that this technology can significantly bring an educational revolution in many educational institutes throughout world. Various research works are being carried out in different universities to extract the benefit of IoT technology in the field of education. The main aim of transforming traditional education into smart education is to enhance the quality of knowledge acquiring process. Dominant focus is on design and implementing IoT-based learning platforms which include smart campus with smart classrooms and laboratories integrated into it. Main focus is on personalized and contextual learning to improve intelligence level of learners, and thereby facilitating their problem solving capability in smart environments. In our chapter, we have discussed various aspects of IoT technology and its significance in educational field. Vital components of smart campus were presented in this study with a simple example for each constituent. Also, we have presented a survey which highlights the impact of IoT technology on students and the manner in which they receive this technology in their day-to-day lives.

References

1. R. Khan, S.U. Khan, R. Zaheer, S. Khan, Future Internet: The Internet of Things Architecture, Possible Applications and Key Challenges, in Proceedings of Frontiers of Information Technology (FIT) (2012), pp. 257–260
2. G. Shen, B. Liu, The visions, technologies, applications and security issues of Internet of Things, in E -Business and E -Government (ICEE) (2011), pp. 1–4
3. L.-y. Zeng, A security framework for internet of things based on 4G communication, in Computer Science and Network Technology (ICCSNT) (2012), pp. 1715–1718
4. R. Abdmeziem, D. Tandjaoui, *Internet of Things: Concept, Building Blocks, Applications and Challenges, Computers and Society*. Cornell University
5. A. Al-fuqaha, M. Guizani, M. Mohammadi, M. Aledhari, and M. Ayyash, *Internet of Things : A Survey on Enabling Technologies, Protocols and Applications*, no. c (2015)
6. N. Gershenfeld, R. Krikorian, D. Cohen, *The Internet of Things*, vol. 291, no. 4 (2004)
7. M. Mohanapriya, *IOT enabled Futurus Smart Campus with Effective E-Learning : i-Campus*, vol 3, no 4 (2016), pp. 81–87
8. H.F. Elyamany, A.H. Alkhairi, IoT-academia architecture: a profound approach, in 2015 IEEE/ACIS 16th International Conference on Software Engineering Artificial Intelligence Network Parallel/Distributed Computer SNPD 2015—Proceedings (2015)
9. J. Chin, V. Callaghan, Educational living labs: a novel internet-of-things based approach to teaching and research, in Proceedings of the 9th International Conference on Intelligence Environment IE 2013 (2013), pp. 92–99
10. H. Cheng, W. Liao, Establishing an lifelong learning environment using IOT and learning analytics, in Advanced Communication Technology (2012), pp. 1178–1183
11. A.P. Castellani, N. Bui, P. Casari, M. Rossi, Z. Shelby, M. Zorzi, Architecture and protocols for the internet of things: a case study, in Pervasive Computing and Communications Workshops (PERCOM) (2010)
12. Y. Chen, X. Dong, The Development and Prospect of New Technology in Modern distance education, Int. Conf. Inf. Sci. Comput. Appl. (2013), pp. 40–44
13. H.F. Elyamany, A.H. Alkhairi, IoT-academia architecture: a profound approach, in 2015 IEEE/ACIS 16th International Conference Software Engineering Artificial Intelligence Network Parallel/Distributed Comput. SNPD 2015—Proceedings (2015)
14. J. Chin, V. Callaghan, Educational living labs: a novel internet-of-things based approach to teaching and research, in Proceedings—9th International Conference on Intelligence Environment IE 2013 (2013), pp. 92–99
15. Y. Wang, Characteristics of English interactive teaching model which based upon internet of things keywords-internet of things; english. Int. Conf. Comput. Appl. Syst. Model. **13**, 587–590 (2010)
16. G. Kortuem, A. Bandara, N. Smith, M. Richards, M. Petre, Educating the internet-of-things generation. Computer **46**(2), 53–61 (2013)
17. Sciforce, Internet of Things for the Classroom. https://www.iotforall.com/internet-of-things-classroom/. 9 April 2019
18. Tom_Bradicich, HPE IoT Innovation Labs: get started with your IoT or Intelligent Edge proof of concept (PoC). https://community.hpe.com/t5/IoT-at-the-Edge/bg-p/internetofthingssolutions/label-name/iot%20lab#.Xg-ZyUczY2w (2018)
19. Nick Flaherty, Advantech updates European IoT labs. https://www.eenewspower.com/news/advantech-updates-european-iot-labs (2017)

A Case for Unikernels in IoT: Enhancing Security and Performance

Siddharth Choudhuri

1 Introduction

IoT systems are on the rise and ubiquitous. While broadly categorized as IoT, such systems encompass sensors based on tiny micro-controllers to more sophisticated processors such as ARM, MIPS, RISCV, or even x86. Of late, the trend of such systems is moving towards having more sophisticated processors than tiny micro-controller based systems. The reasons for such a shift are—(i) the demand to build intelligent systems as opposed to just sensors nodes such as Mote running TinyOS [1]. (ii) the lowering cost of hardware enabling yesteryear's desktop class systems down to few dollars (Eg: Raspberry PI [2] in mass production), (iii) wide variety of choice in IoT hardware that can communicate (eg: Zigbee, 802.xx, ...) with other IoT hardware and/or edge enabling a network of IoT system, and (iv) a data/communication infrastructure availability (eg: 4G) that lets these systems be deployed in locations that was earlier not possible such as remote solar farms, moving vehicles etc. Such data/communication infrastructure also enables the bandwidth required for a mesh of IoT sensors to communicate.

The above factors are driving innovation in this field wherein IoT systems are being deployed both in existing areas for efficiency (e.g.: retail, warehouse, industrial complex) as well as emerging markets that did not exist (eg: car-to-car networks, trucking logistics, ...).

S. Choudhuri (✉)
Irvine, CA 92612, USA
e-mail: siddharthc@gmail.com

2 Motivation

As IoT gets widely used in our day to day lives, security of such systems is of paramount importance. Insecure IoT systems can cause damages anywhere from simple malfunction to taking down critical infrastructure and dangers to human life [3]. While there has been rapid progress in the area of IoT (hardware and software), Operating Systems (OS) aspect of IoT has been either provided by vendors as a reference or platform specific versions of Linux kernel is adopted, both in development and production environment. A choice of existing OS leads to faster time to market and deployment but not necessarily security. The reasons are manyfold including but not limited to:

1. Attack surface: Simply based on the OS image footprint, traditional OS kernel and systems services expose an attack surface that is typically larger than the end application. Even an embedded OS today can span multi million lines of code given the kernel, drivers, and user space system services that are offered. The attack surface can come from kernel (unused drivers, vulnerabilities), user space system services (default services that are left running, open ports, security configurations, ...). The attack surface of a Unikernel is much smaller due to reduced binary (code base).
2. Network Settings: A large class of IoT systems are designed to be deployed in public networks (eg: monitoring utilities, fleet management) and are therefore exposed to a network that is typically not secured unlike private, enterprise networks. Such public networks gives anyone access to monitor network packets, open ports etc, thereby increasing the chances of a compromise if the network is not well secured. Unikernels are custom built and therefore do not include open ports that are not required.
3. Multiple Services and Settings: A traditional OS such as Linux, has many services/daemons running. It is possible that a production IoT system "inherits" these systems from a development environment and default settings of the OS. This is common place when a reference root file system is used which is generic in order to allow for wide use as opposed to an IoT system which needs a very "specific" root file system with just the minimal services enabled. An elaborate service and settings is non-existent in Unikernels.
4. Complex systems to reason: Given the large code base of a traditional kernel like Linux, it becomes extremely hard to reason around security. Even if one starts to configure the kernel, there are well over 100s of modules and services that can need to be first understood and then mapped to make sure that the dependency between modules is clear before removing any. Due to lower code base it is easier to reason about code in a Unikernel.
5. OS Upgrades: While the upgrades on enterprise OS are well defined processes for on-prem data center and cloud, the same for IoT devices on field is still evolving. There is also risk of image bloat with each upgrade. Image bloat due to unwanted modules does not exist in Unikernels.
6. Security Certification: This is an area that is not as evolved as enterprise OS.

An alternative is to avoid the above reasons and use Unikernels instead. Besides security, Unikernels provide other advantages such as smaller image, much faster boot time, and better performance. In the rest of the paper, the case for Unikernels in IoT is described.

3 Unikernels

Unikernels [4–8] are a class of OS where the kernel is compiled into the application. As a result, an application boots when the system starts and eventually runs the application. Therefore, in a Unikernel based system, there is a single address space that is shared between the OS and the application. Further, the OS components are built as a library [8] and therefore an application designer can chose to include which specific components of the OS are included in the Unikernel image. It is assumed that a language runtime support exists for a Unikernel unlike traditional OS. The figure below (Fig. 1) illustrates the a Unikernel vs. traditional OS. By reducing the amount of code, Unikernels can provide enhanced security due to:

- Reduced code: Implies reduced attack surface in the binary.
- The library components and the application can be easily verified (whether formally or not) compared to a traditional OS where it is almost impossible to verify given the interleaving code and dependencies.

Since an application and OS components (provided as library functions) are compiled into a single image, by definition, Unikernels are specialized in the sense that they can only run the "application" unlike a multi-programming environment. Note

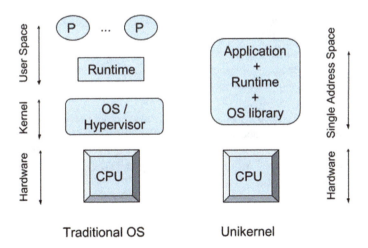

Fig. 1 Traditional versus Unikernel

that IoT systems are domain specific systems and do not run a generic, arbitrary code, they fit well in the class of Unikernel OS.

In the next section, a case for Unikernels for IoT is presented. Finally, drawbacks of using Unikernels is briefly presented. However, the advantages overweight the disadvantages.

4 Case for IoT Unikernels

This section provides a case of using Unikernels for IoT illustrating the advantages of such an approach. While security is a major advantage, there are other features of Unikernels that align well with IoT systems.

4.1 Security

One of the major advantage of using a Unikernel is security. Kernel (OS) level security vulnerabilities can often led to catastrophic system compromise (eg: root access, ability to install malicious driver, etc.). A Unikernel for IoT is advantageous from a security point of view due to both smaller attack surface attributed to less code and the possibility that the small code base can be formally verified. Note that the assumption here is that of a smaller code base which is reasonable and practical for an IoT system. IoT systems are not general purpose and therefore including the required drivers for sensors, actuators and minimal library OS functionality (storage, networking) can lead to a Unikernel whose library OS component can be reasoned about and verified.

Besides the smaller attack surface, the system services expected out of a Unikernel is pretty much a network service that can serve mostly to transmit sensor data and receive commands from either the edge or the cloud. Note that this greatly simplifies by not having any daemons/services that run by default in a traditional OS. Consequently, the IoT system exposes hardly any ports (sockets) other than a network send/receive.

The third aspect of security is a kernel upgrade. Traditional OS has to be periodically updated based on security vulnerabilities discovered by OS vendors (Eg: heartbleed vulnerability [9]). The lower code base in a Unikernel greatly reduces this risk. Also, with a Unikernel for IoT the periodicity and size, if any, of such upgrade will be less frequent compared to traditional OS.

4.2 Performance

Performance can be characterized in many different ways depending on the application needs. The natural advantages of using a Unikernel benefits IoT systems as well. Since the application and OS are fused together, running the application "boots" the system. Therefore, an application can be up and running in a matter of couple of seconds as opposed to 10s of seconds required to boot and embedded OS. This makes the IoT system feel more like a device that turns on as opposed to a server that takes a while to boot and thereafter is ready for its main purpose.

An IoT Unikernel reduced foot print will result in lower storage space. This is especially useful in IoT systems that typically rely on flash/Non-volatile memory as secondary storage. The drastic reduction in OS footprint can be turned into an advantage either in terms of having lower size flash (i.e., cost reduction) or using the excess storage for longer logs etc.

Since an IoT Unikernel runs in a single address space, this also has positive implications on the RAM requirements of the IoT system. Lowering RAM can greatly improve the power consumption of the IoT system. Also, RAM typically has higher $cost/\$$ and therefore lower RAM can lead to lower TCO for IoT system. Just like storage, savings in RAM can also be used by the application to either cache or help with latency.

4.3 Software Engineering

Using a Unikernel also helps in simplifying the development and production workflows. Typically a build system involves compiling the library OS and application into a single binary. Such a binary can be either run on the actual IoT system or an emulation system (e.g.: qemu [10]). The typical ramp up for developers only involves the code flow from boot to the application without worrying about details of the OS that are not included in the Unikernel. Besides this, the lowered complexity of Unikernels can result in faster ramp up of developers.

5 Challenges

While this paper makes a case for the use of Unikernels for IoT systems, there are challenges that are described below:

1. Language runtime support is a prerequisite for using Unikernels. Give a language of choice that an application is written in, there has to be a support for the same language runtime in Unikernel. (For example, IncludeOS for C++, OSv for C, ...).

2. Single address space also means that the memory errors can lead to system crash as opposed to application crash in a traditional OS.
3. Limited debugging capabilities compared to mature debugging tools (Eg: gdb, valgrind, gprof ...) for a traditional OS.
4. Unikernels are making a comeback due to cloud and containers. Therefore, this is still a newer field compared to traditional OS although ideas such as Exokernels [8] were proposed in the early 90s.
5. There are specialized kernels for IoT such as TockOS that can provide a balance between a Unikernel and traditional OS [11].

6 Conclusion

Unikernels properties have several similarities with that of IoT—more secure, lower code footprint, subsecond boot times, application specific use to name a few. Besides these software similarities, Unikernels lend themselves nicely to hardware constraints (RAM, power, storage ...) that IoT systems have. This paper argues that the merits of a Unikernel OS lends itself nicely to IoT and therefore, IoT systems should be developed on Unikernels as opposed to general purpose OS that are server grade and have been retrofitted to be used in IoT. This paper emphasizes the higher security and better performance of Unikernels. Finally, the challenges of using Unikernels in IoT are enumerated.

References

1. O.S. Tiny, P. Levis, S. Madden, J. Polastre, R. Szewczyk, A. Woo, D. Gay, J. Hill, M. Welsh, E. Brewer, D. Culler *TinyOS: An operating System for Sensor Networks* (Springer, Berlin, 2004)
2. Raspberry Pi. https://www.raspberrypi.org/
3. Hackers remotely kill a Jeep on highway. https://www.wired.com/2015/07/hackers-remotely-kill-jeep-highway/
4. A. Madhavapeddy, D.J. Scott, Unikernels: Rise of the Virtual Library Operating System. Queue 11, 11, Pages 30 (December 2013), 15 p (2013). https://doi.org/10.1145/2557963.2566628
5. A. Madhavapeddy, D.J. Scott, Unikernels: the rise of the virtual library operating system. Commun. ACM **57**(1), 61–69 (2014). https://doi.org/10.1145/2541883.2541895
6. A. Bratterud, A.-A. Walla, H. Haugerud, P.E. Engelstad, K. Begnum, IncludeOS: A minimal, resource efficient Unikernel for cloud services, in *2015 IEEE 7th International Conference on Cloud Computing Technology and Science (CloudCom)*
7. R. Pavelick *Unikernels* (O'Reilly Media, Inc.) https://www.oreilly.com/library/view/Unikernels/9781492042815/
8. D.R. Engler, M.F. Kaashoek, J. O'Toole, Jr, Exokernel: an operating system architecture for application-level resource management, in *Proceedings of the Fifteenth ACM Symposium on Operating Systems Principles (SOSP '95)*, ed. by M.B. Jones (ACM, New York, NY, USA, 1995), pp. 251–266. https://doi.org/10.1145/224056.224076
9. http://heartbleed.com/
10. www.qemu.org

11. A. Levy, B. Campbell, B. Ghena, D.B. Giffin, P. Pannuto, P. Dutta, P. Levis, Multiprogramming a 64 kB computer safely and efficiently, in *Proceedings of SOSP '17* (ACM, New York, NY, USA, 2017)
12. Damage IoT bugs can do. https://www.wired.com/story/elaborate-hack-shows-damage-iot-bugs-can-do/

Cloud of Things Assimilation with Cyber Physical System: A Review

Yashwant Singh Patel, Manoj Kumar Mishra,
Bhabani Shankar Prasad Mishra, and Rajiv Misra

1 Introduction

In current scenario the cloud and IoT are two very distinguished interoperable information and communication technologies widely adopted in a different application of resources for the client application, with 'Pay As You Go (PAYG)' utility computing model. It becomes quite successful in several application areas such as resource virtualization [1], remote processing [2], on-demand data centers [3], big data analytics [4], large scale data processing and services to end-users. While the IoT paradigm has stimulated from being the futuristic vision to a certain degree of market realism. Major ICT key players like Cisco, Apple, and Google focused themselves on building the number of connected objects in their network. Integration of cloud of things and cyber-physical devices broaden the performance of future Internet to execute the transmission of digital data for controlling physical data generated from such cyber-physical devices. Recently, CPS has been extensively followed in the application areas such as industry, intelligent highways, environmental monitoring, aerospace systems, military applications, robotic system, process control, agriculture, home monitoring, and others. The researchers have conducted an extensive study to find out the diversity between CPS applications as well as its presented framework to integrate it with some other existing concepts and technologies. Figure 1 shows a brief report on Google Search Trends for terms Cloud, IoT, CPSs, Cloud of Things and CPS Clouds from 2015–2019. This study is performed based on the publications

---

Y. S. Patel (✉) · R. Misra
Department of Computer Science & Engineering, Indian Institute of Technology Patna, Patna, India
e-mail: yashwant.patelasct@gmail.com

M. K. Mishra · B. S. P. Mishra
School of Computer Engineering, Kalinga Institute of Industrial Technology Deemed to be University, Bhubaneswar, Odisha, India

© The Author(s), under exclusive license to Springer Nature Singapore Pte Ltd. 2021
S. Kumar Pani and M. Pandey (eds.), *Internet of Things: Enabling Technologies, Security and Social Implications*, Services and Business Process Reengineering,
https://doi.org/10.1007/978-981-15-8621-7_8

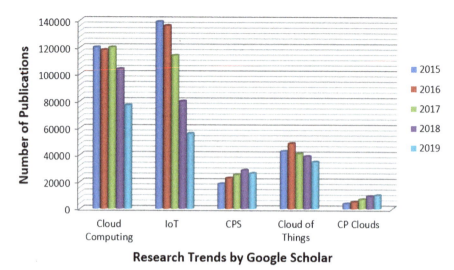

Fig. 1 Google search trends for terms cloud, IoT, CPSs, cloud of things and CPS clouds (*Sources* Google Scholar, DBLP)

in the respective fields. It shows the major growth in the trends of Cloud, IoT and CPS concepts. However their integrations are still found to be in a preliminary phase.

The 3 key objectives to integrate CPS with the CoT paradigm are as follows: First, CPS devices are very much resource-constrained and also have limited storage and processing competence. Secondly, the CPS devices generate a massive amount of continues data. Third, the lack of objects and standard interfaces to make their heterogeneous devices interoperable on the Internet. In terms of framework, CPS and IoT are two different levels of digital integration. At one hand, the CPS defines the seamless integration between the physical and digital (cyber) worlds. On the other hand, IoT is simply the core networking of these 'Cyber-Physical Things' to transfer the information. The CPS defines the first level, and IoT represents the second level of digital integration. Since these two domains are gradually realized around the technology globe, a new integration of Cloud computing is introduced at the Third level. It will not only provide an abundance of resources but also ease the limitless computations.

In the latter part of this chapter, we first give a holistic view on cloud of things paradigm and discussed the speedy development in the area of Cloud Computing, IoT, SOTs, and cyber socialization for large scale computation. Next, in Sect. 3, CPSs are elaborated in a nutshell. We mainly focused on architectural view of CPS using the similar concepts and also discuss the domain specific applications of CPS. Then, we present a layered architecture on a cloud with the integration of IoT and CPS in Sect. 4. Section 5 gives the glimpses of research challenges while adopting integration. A brief case study of cloud of things paradigm with CPS and its applicability scenario in cyber-physical disaster environment is illustrated in Sect. 6. At last, the concluding remarks are presented in Sect. 7.

2 Holistic View on Cloud of Things

To handle large scale computation, there is a speedy development in the area of Cloud Computing, IoT, SoTs, and cyber socialization. Due to the rapid increase in connected IoT devices, massive volume of data are being produced every minute. Storage and processing of such huge volumes of data require additional rental storage space with primary goals of high performance and efficient resource utilization. All this is possible with the IoT and Cloud integration, also termed as Cloud of Things (CoT) paradigm. The IoT and cloud computing are two distinct models having independent advantages. On the one hand, the cloud can leverage better economies of scale and performance with unlimited rental storage and high resource utilization. On the other hand, the IoT solves the real-life problems using lightweight solutions with minimal cost in a more distributed and dynamic manner. The key harmony between both the terminologies are shown in Table 1.

2.1 Applications of Cloud of Things

Cloud of Things has several impacts on smart applications like health care, smart city, etc. as depicted in Fig. 2. This section gives a brief glimpse of a rich variety of applications generated with the integration of Cloud-IoT paradigm.

1. Healthcare: Minimize the expertise requirements and produce the dominating solutions for efficient management of health care data.
2. Smart City: Provide future-oriented services to acquire information by accessing geo-locations.
3. Smart Home: Create automation of smart home devices.

Table 1 Harmonizing cloud and IoT paradigm

Cloud computing	IoT
Manage big data	Source of big data
Virtualized resources	Real-world things
Ubiquitous computing	Pervasive computing
Unlimited computational power	Limited computational power
Virtually unlimited storage	Limited storage
Internet for delivery of services	Internet for enabling convergence point

Fig. 2 Applications of cloud-IoT

4. Video Surveillance: To provide efficient storage, management, and extraction of knowledge from IP *Internet Protocol) cameras (i.e., network cameras).
5. Smart Grid and Smart Energy: To provide self-healing, energy distribution, and quality management.
6. Automotive and Smart Mobility: To minimize road congestion, enhance road safety, vehicle recommendation, and traffic management with the help of Satellite Network, Wireless Networks, and RFID, etc.
7. Environment Monitoring: Provide continuous monitoring of soil, water, landslides, civil structures, food quality, and heat radiations etc.
8. Smart Logistics: To automate the flow control of goods and services from source to destination by using geo-locating technology.

2.2 Intelligent Choice for Internet of Things

In 2015, 75% of the world's population (7.3 Billion) was connected to the Internet, increasing from 15% a decade before 2005. Hence internet penetration to things

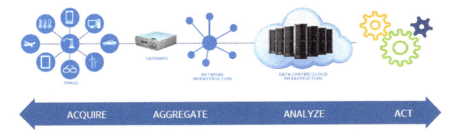

Fig. 3 Typical IoT ecosystem

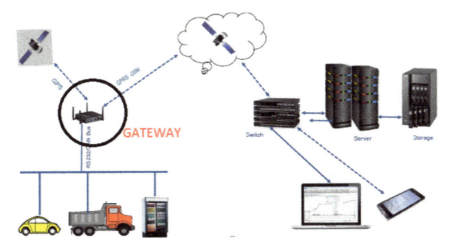

Fig. 4 Importance of gateway

inevitable. So it is crucial to understand how things connect to the internet (Cloud). Typical Internet of things ecosystem involves below items as shown in Fig. 3.

If IoT is characterized as connecting things with the internet, one can visualize the importance of gateway from Fig. 4. A gateway is a network point that acts as an entrance to another network. Another widespread usage of this term is the payment gateway, which accelerates the information exchange between a payment portal i.e., smartphone, website etc., front-end processor or collecting from financial institution. In a similar context, Gateway is used in IoT ecosystem. In the IoT context, the gateway is a device which can connect to things on one side and connect to Internet Cloud on the other side. The gateway supports protocols such as Rs232, Rs485, Modbus, BacNet, DIO, ADC, SPI, TTL. The gateway supports the connectivity to cloud on GSM, LAN (ethernet 10/100), Zigbee, Zwave, Bluetooth, Wifi.

An ideal gateway has the following characteristics:

1. Connectivity—Wide variety of options
2. Manageability—Easy and quickly able to manage the devices
3. Security—Foolproof security.

Table 2 Difference between CISC and RISC

CISC architectural design	RISC architectural design
More emphasis on hardware for building complex instructions	More emphasis on software than hardware
More addressing modes	Fewer addressing modes
Supports multiple types of instruction sizes and formats	Supports instructions of same set with lesser formats
Uses lesser number of registers	Uses more number of registers
Vast use of microprogramming	Complexity in compiler
Difficult to implement pipelining	Easy to implement pipelining
Instructions consume a varying amount of cycle-time	Instructions consume one cycle-time

There are several popular gateways available in IoT such as Digi, Cisco, Libellium, MediaTek, Freescale, National Instruments, Texas Instruments, Intel.

Current processors fall under two classification: first is complex instruction set computing (CISC) processor and the other one is reduced instruction set computing (RISC) processor. In simpler terms, CISC caters/supports HLL (High-level languages) with complicated instructions while RISC applies a large number of registers, single CPI (cycle per instruction) and pipelining. The key differences between CISC and RISC processors are presented in Table 2.

Performance of a processor is defined as follows:

$$\text{Performance} = \text{Time/program} \qquad (1)$$

Let us rewrite this equation:

$$\frac{\text{Time}}{\text{Program}} = \frac{\text{Time}}{\text{Cycle}} * \frac{\text{Cycle}}{\text{Instruction}} * \frac{\text{Instructions}}{\text{Program}} \qquad (2)$$

So there are two ways to achieve better performance—Either minimize the instructions or reducing the cycles.

CISC follows minimizing the instructions (sacrificing the number of cycles per instruction). RISC applies to reduce the cycles per instructions ignoring the number of instructions per program. ARM holding came up with RISC based architecture.

ARM stands for Advanced RISC Machine. ARM architecture works better under low power and capable of having a smaller size. Having a complex set of instructions to have better performance is required for IoT. Now, most of the gateways mentioned above follow ARM-based chips (i.e., RISC ISA).

In the IoT world, below are major criteria to be considered than Size, power of the mobile world. They are Security, Instant BSP support, Inbuilt OS, massive ODM support. Before making the comparison, among the gateways, it is found from the literature that Raspberry Pi as Gateway is the best and cheapest option. But RasPi is

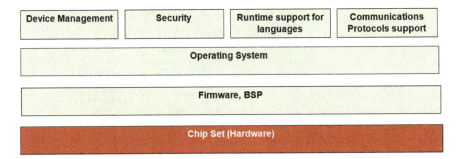

Fig. 5 Basic components of gateway

not for production systems, at the maximum, raspi model can be used for Proof of concept due to the following reasons. Firstly, it has a compliance problem. Secondly Boxing of the Raspi itself a difficult task considering the heat generated by Raspi systems. Thirdly, production issues, it is not possible to mass produce the raspi based models. As Raspi is a development environment, even licensing will be a problem for industrial production systems. Gateway is associated with four crucial aspects such as Device management, Security, Runtime support for languages, and openness and Communications channels with both things and cloud. The gateway consists of several essential components as presented in Fig. 5.

The components are summarized as follows:

1. Chipset: It is the hardware layer.
2. BSP: All Intel gateways come with pre-loaded BSPs Thus, plug the power, almost you' re up. This is a core strength. Again with strong background, successful history in PC and server market will help Intel to have better BSPs
3. OS: Intel had acquired wind river OS. Then working on fully open source OS, wind river offers better support. WR is an already proven OS.
4. Device management: Intel Gateways support OMA DM and TR69 complaint server components and fully functional client components readily built-in gateway.
5. Security: Intel had acquired McAfee, a premier security company which is popularly known in the PC world, can address all IoT security issues.
6. Runtime support: Unlike other gateways, Intel offers an open environment to build applications in Java/Lua/Python and many other languages.
7. Communication Protocols: Based on need, communication channels can be arranged

Thus Intel offers better support in all aspects of the gateway as against other boards. Summary of the mentioned points is represented in Fig. 6.

Ease of development: Using IDP (intelligent device platform), one can build apps quickly. This IDE is built on the open source Eclipse framework.

ODM support: Device manufacturers support is the key for Intel because Intel is in the same business with these manufacturers for the PC market. So it is easy for

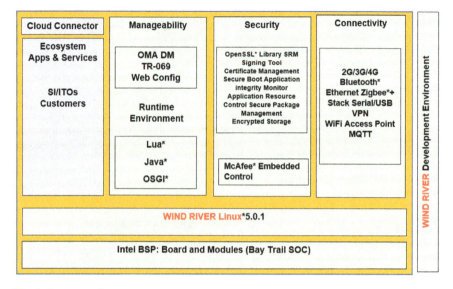

Fig. 6 Summary of gateway

them to identify reliable ODM partners quickly. Aaeon, Adlink, EuroTech, Kontran, NexCom, Vantron, AdvanTech, Vantron are few known ODM partners of Intel.

3 Application Based Research Challenges in Cloud of Things

The research opportunities involved with the cloud of things will create several research challenges as well as more significant security threats. The integration of heterogeneous networks, services, and data, will bring numerous research challenges with different applications, as discussed in Fig. 7.

4 Cyber Physical Systems in a Nutshell

The 21st-century digital electronics have brought explosive growth in computing power, communication technologies and artificial intelligence (AI) capabilities that together have contributed in the linking of cyber (digital) systems having sensors and actuators with the physical environment what is known as cyber-physical systems (CPS) [5]. Modern CPS also inspires a novel vision of the future computing, where the monitoring of real-world is done via smart sensors that transfer sensing information into the cyber-world, where the ubiquitous apps and services make use of the collected data to reflect the physical environment in real-time for enhancing reliability, efficiency, functionality and security etc. In 2006, the CPS phrase firstly introduced at the National Science Foundation (NSF) [6]. Since then, the NSF has increasingly sanctioned several grants and opportunities to the research community to encourage innovative research projects on CPSs. The technology of CPS much depends on the embedded systems, communication mode, and integrated software to the devices. CPS gives another dimension to humans while interacting and controlling the physical world. Applications of CPS involved in different domains such as defense systems, healthcare, aerospace, robotics, transportation, process control, manufacturing, tele-physical operations, disaster management, agriculture, large-scale infrastructure, smart grid, and society, etc. Several articles have been presented on CPS to describe its system-level design, features, major challenges, and applications, etc. Shi [7] describes an overview of CPS characteristics, challenges, and applications. Baheti and Gill [8] introduced the CPS model and further research directions on its design. Horvath and Gerritsen [9] describe on CPS designs, principles, and features. Rajkumar [10] written on system-level aspects of CPS. Lee [11] recommended two methodologies, that is physicalizing the cyber (PtC) and cyberizing the physical (CtP) to integrate the cyber systems with the physical systems.

4.1 Architectural View of CPS

The CPS design is integrated with the physical world applications, i.e., for monitoring and controlling, cyber systems, i.e., embedded devices for processing and communicating with their distributed computing environment and interfaces, i.e., other intermediate components and communication network as shown in Fig. 8. CPS

	Research Challenges	Privacy	Security	Scaling	Reliability	Extensibility	Availability	Efficiency	Energy Saving	Cost Effectiveness	Mobility	Maintainability	Performance	Legibility
Applications	Healthcare		✓	✓			✓	✓		✓	✓	✓	✓	✓
	Smart City			✓	✓			✓	✓	✓	✓		✓	✓
	Smart Home			✓	✓			✓	✓	✓	✓		✓	✓
	Video Surveillance	✓	✓			✓		✓					✓	✓
	Smart Energy and Smart Grid			✓	✓			✓	✓	✓	✓		✓	✓
	Automotive and Smart Mobility	✓				✓	✓		✓	✓	✓		✓	✓
	Environment Monitoring	✓	✓			✓	✓						✓	✓
	Smart Logistics							✓		✓			✓	✓

Fig. 7 Research challenges for cloud of things adaptation

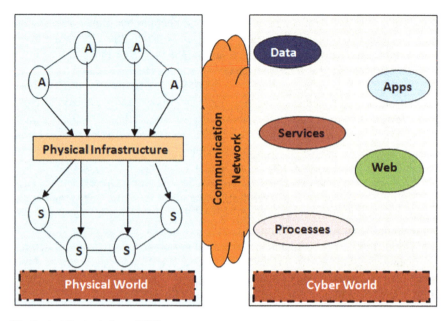

Fig. 8 Architectural view of CPS

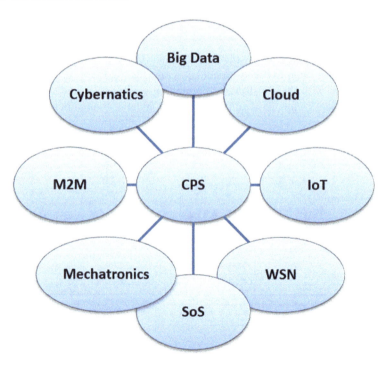

Fig. 9 Similar concept of CPS

Table 3 Applications of CPS

Domain	Function
Military	To develop intelligent and autonomous complex systems
Robotics	Offer services for the benefits of humans and remotely control the robots
Biomedical and Healthcare	Design and monitor the health care systems
Transportation	Traffic management of aircraft systems
Manufacturing	Productivity of goods and services
Disaster Management	To handle the disaster and protect the infrastructures
Agriculture	To make smarter and efficient agricultural systems
Energy	To make energy efficient infrastructures
Aerospace	To be operated in highly dynamic environment
Society	To design special purpose Apps for task automation via hardware devices

contains a range of similar concepts and terminologies, as shown in Fig. 9 such as (i) Big Data, i.e., to investigate and manage the large and complex data sets, (ii) Cloud i.e, provide internet on-demand network based access for shared pool of configurable computing resources, (iii) Cybernetics i.e. to study the control and communication features of living beings, organizations and machines, (iv) IoT i.e. idea of communicating smart devices through radio frequency identification (RFID) tags, (v) Web of Things i.e. to integrate the objects of real-world to the internet via leading web technologies. (vi) Systems of Systems (SoS), i.e., large-scale systems to achieve the community goals and (vii) Mechatronics, i.e., to integrate the mechanical and electronic systems. M2M i.e. Machine to Machine communication.

4.2 Applications of CPS

This section summarizes the domain-specific application of CPS. Table 3 provides an outline of CPS applications.

4.3 Research Challenges of CPS

This next generation physically aware technology is also associated with some major research challenges as described in Table 4 [12].

Table 4 Research challenges of CPS

Research challenge	Description
Standards	To adopt standards among heterogeneous systems
Privacy and Security	To deal with security methods and privacy concerns
Real-time	For time-critical data delivery
Abstraction	Capabilities to work with the large-scale data
Reliability	For maintaining the degree of accuracy
Robustness	To operate during failures

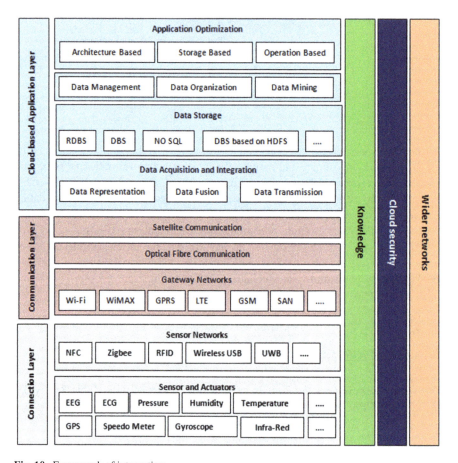

Fig. 10 Framework of integration

5 Framework of Integration

The integration framework is a typical 3-layered architecture as referred from [13, 14]. It consists of the collection, communication, and application layer, as represented in Fig. 10. The detailed study of layered functional modules is described as follows:

1. Connection Layer: It manages the gathering of sensor-generated data from the variety of source like EEG, ECG, and GPS, etc. and enables them to be connected with sensor networks like Zigbee, Wireless USM, NFC or wide area networks, etc. The sensor networks transfer data to the gateway networks.
2. Communication Layer: At communication layer, gateway networks such as Wi-Fi. WiMAX, GPRS, and GSM, etc. transfer data to cloud-based platforms for storage and cloud services.
3. Cloud-based application layer: Application layer enable the technologies for intelligent information processing and analysis with the help of middlewares and business models. The cloud services are designed for three categories of users, i.e., provider, operator, and consumer. It processes the applications based on their demand. At this layer, various modules are discussed as follows:

    (a) Data Representation: It is used for data acquisition and integration. Multimedia data gathered from different sensor devices is converted into a flexible and common format, i.e., XML encoding for sensor attributes and measurement.
    (b) Data Fusion: Heterogeneous data acquired from various sensors are merged to create a common view for future operation. This process is called a fusion of multi-source data.
    (c) Data Transmission: After the process of data acquisition, data is transmitted to the back-ends. UDP protocol is commonly used for transmission of multimedia data.
    (d) Data Storage: Sensor generated data consume large storage space. The storage types of data can be classified into RDBMS, DBMS, NoSQL, and HDFS, etc. To support various types of data, all should be combined for effective data storage.
    (e) Data Management: It forms an effective data based for intelligent processing of distributed or parallel applications. The process can be categorized into the management of metadata, semantic annotation, and indexing, etc. For distributed processing of data, various popular processing methods are applied for parallelization, fault-tolerance, and scalability, etc. While the data mining module is designed for efficient and effective data processing.
    (f) Application Optimization: At the application layer, the optimization modules are viewed as: Architecture based: caching-based, message-oriented and middle-ware based, etc. Storage-based: Hierarchical extended storage, Removal of duplicate-data, File-grouping and Buffering of I/O requests, etc. Operation based: Search space minimization, Priority-based scheduling, and nonintrusive slot layering, etc.

6 Future Challenges with Cloud for Management of IoT and CPS Generated Data

The key research challenges involved with cloud for management of IoT and CPS generated data are:

1. Selection of Architecture and standards
2. Connection technologies and physical manufacturing of resource abilities
3. Virtual resource management
4. Modelling of cloud services
5. Isolation and Sharing
6. Security
7. Storage and Scalability
8. Service Composition
9. Service Integration and Management

7 Case Study on Applicability of Cloud of Things Paradigm for Cyber-Physical Disaster Management Systems

Mitigation actions are necessary to minimize the risk to life, property, social, and natural resources from natural and man-made hazards. With the extensive application of smart devices, such as, sensors, home appliances, smart-phones, Internet of Things (IoT), tablets, laptops, and even cars that consist of CPS and their ubiquitous network access, and multimedia content sharing to the cloud provide a non-trivial opportunity to share information of disaster regions. Such smart devices will endlessly cause a massive volume of data than any personal web app. The connected cyber universe including organizations reside escalating in giant data centers linked to billions IoT connected devices, all supervised via increasingly intelligent software. For handling such independent requirements, the cloud delivers services through geographically distributed data centers across the world, which can host small numbers to thousands of servers. Cloud providers have implemented various open-source software based distributed database systems for storing, processing, and simplify the management of huge volume of workloads across clusters of commodity servers. For knowledge extraction from the collected data and feeding end-users across smart city applications and sectors, such system follows a classical 3-layered approach, as shown in Fig. 11.

At level-1, the collection layer collects data via distributed smart devices and transfer to the gateways. At level-2, the transmission layer transfers data from gateways to distributed cloud-based platforms. Finally, at the processing layer, data is processed in the cloud-based framework where the knowledge extraction performed and also provided to the applications. In the course of the dynamic resource allocation of CPS and IoT related application workloads, the different types of virtualization

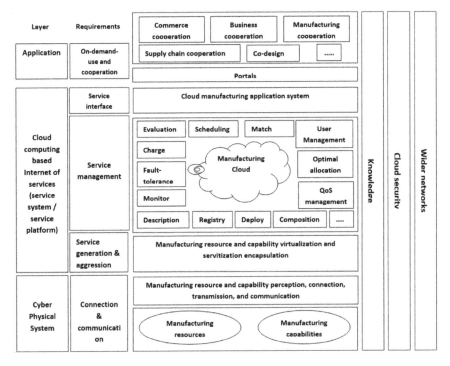

Fig. 11 Layered architecture of cloud based cyber physical system services based on [13, 14]

technologies, i.e., on-demand and advance resource provisioning not only make the cloud services affordable for the cyber-physical-disaster system but also efficiently reduces energy-consumption. The applicability of the virtualization technologies in the disaster management situation is described as follows:

1. The multi-tenant architecture of the cloud-enabled cyber physical systems composed of three major components: multi-tenant cloud data, application and user interface as represented in Fig. 12. This model can scale massive information in cloud of each site.
2. In crowded geographical regions, use cases such as railway stations, school zones, smart buildings, shopping malls and traffic zones where the user access internet using WiFi networks comprising of thousands of access points. The access point is an intelligent edge device that continuously keeps track of user details through connectivity. Such information of activities acquired through cyber physical systems are stored in multi-tenant cloud.
3. Every access point contains within its coverage region periodically shares summarized information to the multi-tenant cloud architecture for further analysis and correlation as well as real-time and historical connection logs.

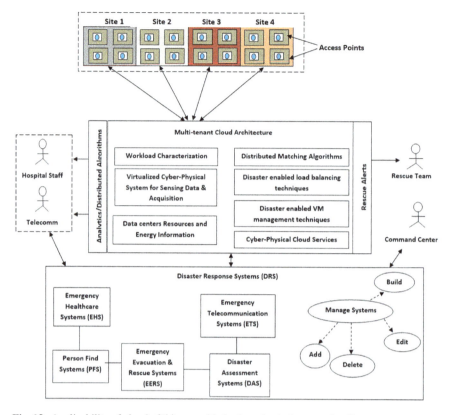

Fig. 12 Applicability of cloud of things enabled cyber-physical system for disaster management

4. In multi-tenant cloud architecture, a relationship map of disaster management application, analytics, access points and alerts is created as that data is pushed to the cloud.
5. In cloud, the massive collected data through different sensors will be analyzed and transformed as necessary feedback or valuable actionable information to people in the physical world.
6. For example, the citizens inside the disaster zone may receive the information of evacuation route or the navigation can also enable the users to move through an app by analyzing their local surrounding environment.
7. The architecture uses distributed matching algorithms on multi-tenant cloud to match details (such as-disaster-event location maps, disaster affected people etc.) to the rescue teams. This information will help the rescue teams to assess the overall damage and loss.
8. To transfer this data amid systems and services, the substantial heterogeneous cloud-enabled cyber-physical services such as medical devices, smart alarms, health information systems, satellite, call centers and social network services etc. inside the framework give rise to volume and variety of data which is collected,

integrated and interpreted for performing targeting actions and building up-to-date knowledge for the disaster response systems such as disaster assessment, healthcare, telecommunication, and emergency evacuation and rescue systems etc.

9. The dashboard user interface is used to display the complete analysis in the context of live failures and relevant historical logs to help speed up troubleshooting efforts.
10. To save human life during disaster, the cloud based system accurately maps the details of users.

8 Conclusion

With the substantial growth in the IoT-based deployments and cloud computing technologies, a rich set of applications have been founded in the real world environment. The IoT has made the technologies smarter and intelligent by allowing communications between objects and human through the internet. While on the other hand, the emerging domain of CPS has attracted the researcher's interest in the field of energy control, model-based software design, and control techniques. The concept behind CPS is to concentrate on the coupled system design between cyber and physical world. To shed the lights on IoT and CPS, this chapter firstly gives a brief study on concepts, applications, and research challenges. Then an underlying layered architecture is introduced to deliver an extensive interpretation of the involvement of cloud with IoT and CPS. This study gives a brief overview of the integration of cutting-edge technologies and how this integration will create system-level challenges for diverse applications. On this basis, several research issues are listed for the cloud with the management of IoT and CPS generated data. To understand the applicability of cloud of things paradigm for CPSs, a real-world scenario is also discussed.

Acknowledgements We thank the anonymous reviewers and the editor of ICDCIT industry symposium for their expertise comments and valuable suggestions, which have helped us to improve the quality and acceptability of the chapter significantly. It is acknowledged that the work by Y.S. Patel is partially supported by Department of Science & Technology (DST), Govt. of India, New Delhi, India under ICPS Programme through the Project Number: T-403, "Low-cost Energy-Efficient Cloud for Cyber-Physical Disaster Management Systems." He also acknowledges Visvesvaraya Ph.D. Scheme for Electronics and IT, an initiative of the Ministry of Electronics and Information Technology (MeitY), Government of India <MEITY-PHD-2525> for support.

References

1. Q. Duan, Y. Yan, A. V. Vasilakos, A survey on service-oriented network virtualization toward convergence of networking and cloud computing. IEEE Trans. Network Service Manage. **9**(4), 373–392 (2012)

2. M.B. Mollah, K.R. Islam, S.S. Islam, Next generation of computing through cloud computing technology, in *25th IEEE Canadian Conference on Electrical and Computer Engineering (CCECE)*, Montreal, QC, pp. 1–6 (2012)
3. J.K. Muppala, M. Hiltunen, R. Stroud, J. Wang, The first international workshop on dependability of clouds, data centers and virtual computing environments, in *IEEE/IFIP 41st International Conference on Dependable Systems & Networks (DSN)*, Hong Kong, pp. 590–591 (2011)
4. C. Ji, Y. Li, W. Qiu, U. Awada, K. Li, Big data processing in cloud computing environments, in *12th International Symposium on Pervasive Systems, Algorithms and Networks*, San Marcos, TX, pp. 17-23 (2012)
5. E.A. Lee, Cyber-physical systems—are computing foundations adequate? in *Position Paper for NSF Workshop on Cyber-Physical Systems: Research Motivation, Techniques and Roadmap*, Austin, Texas, October 16–17 (2006)
6. E.A. Lee, S.A. Seshia *Introduction to Embedded Systems: A Cyber-Physical Systems Approach*, 1st ed (2011)
7. J. Shi, J. Wan, H. Yan, H. Suo, A survey of cyber-physical systems, in *IEEE International Conference on Wireless Communications and Signal Processing* (2011)
8. R. Baheti, H. Gill, Cyber-physical systems, in *The Impact of Control Technology* (IEEE, New York, 2011), pp. 161–166
9. I. Horvath, B.H.M. Gerritsen, Cyber-physical systems: concepts, technologies and implementation principles, in *The Tools and Methods of Competitive Eng. (TMCE) Symposium*, pp. 19-36 (2012)
10. R. Rajkumar, I. Lee, L. Sha, J. Stankovic, Cyber-physical systems: the next computing revolution, *47th IEEE/ACM Design Automation Conference*, pp. 731–736 (2010)
11. E.A. Lee, CPS Foundations, in *47th IEEE/ACM Design Automation Conference* (2010)
12. Rihab Chari, Fatma Ellouze, Anis Kouba, Basit Qureshi, Nuno Pereira, Habib Youssef, Eduardo Tovar, Cyber-physical systems clouds: A survey. Comput. Networks **208**(24), 260–278 (2016)
13. P.P. Ray, A survey on Internet of Things architectures. J. King Saud Univ. - Comput. Inf. Sci. **30**(3), 291–319 (2018)
14. F. Tao, L. Zhang, V.C. Venkatesh, Y. Luo, Y. Cheng, Cloud manufacturing: A computing and service-oriented manufacturing model. Proc. Inst. Mech. Eng. BJ. Eng. Manuf. **225**(10), 1969–1976 (2011)

CPSIA information can be obtained
at www.ICGtesting.com
Printed in the USA
LVHW082047180121
676820LV00001B/5